Popular Science®
Do-It-Yourself
Yearbook 1991

Edited by Al Gutierrez

Popular Science Books

Meredith® Press
New York

Cover Photo by: Jonathan Press

For Meredith® Press:
    Director: Elizabeth P. Rice
    Editorial Project Manager: Connie Schrader
    Assistant: Ruth Weadock
    Production Manager: Bill Rose
    Copy Readers: Daniel Early, Gene Schnaser

For Jonathan Press:
    Produced by: Jonathan Press, Cannon Falls, MN
    Producer/Executive Director: Al Gutierrez
    Copy Editor: Berit Strand
    Copy Editor: Cheryl Clark
    Copy Editor: Barb Machowski
    Technical Consultant: Gary Branson
    Illustrator: Geri Klug
    Woodworking Technician: Pat Manion

Book Design: Jonathan Press

Special thanks to:
    California Redwood Association, Novato, CA
    Century Mfg. Co., Minneapolis, MN
    Georgia-Pacific Corporation, Atlanta, GA
    Minwax Company, Inc., Montvale, NJ
    Sears, Chicago, IL
    Shopsmith, Inc., Dayton, OH
    Workbench magazine, Kansas City, MO

The projects presented in this book have come from a variety of sources. The plans and instructions have been developed by many different writers. We have tried to select those projects that will assure accuracy and safety. Due to differing conditions, tools, and individual woodworking skills, Meredith® Press assumes no responsibility for any injuries suffered, damages or losses incurred during, or as a result of, the construction of these projects.

Before starting on any project, study the plans carefully; read and observe all the safety precautions provided by any tool or equipment manufacturer and follow all accepted safety procedures during construction.

Preface

Thinking of building a new home, making some home improvements, building a tool chest or learning wire welding? Well, you're in luck! This year's *Popular Science® Do-It-Yourself Yearbook* provides plans, ideas and comprehensive construction information on home design and construction (page 1), home remodeling and improvements (page 21), workshop projects (page 77), tool use and techniques (page 151), home and auto management (page 178) and a whole lot more.

I. Home Design & Construction. This section provides worthwhile information about innovative techniques like steel-framed home construction (page 11). With lumber becoming one of our most valuable resources, we need to examine alternative construction materials such as steel.

If you have been looking for practical home application programs for your computer, we have reviewed one. It is entitled *Design Your Own Home: Architecture*. This unique program lets you create and design your own home (page 15). As many times as you modify your dream home, the program alters your schematic design.

II. Remodeling & Improvements. People seem to be working harder and spending more time and money on making their home a showpiece for their own enjoyment. In this section you'll learn how to effectively use lighting to beautify your landscape (page 46), how to make practical home improvements like adding a bay window (page 30) or how to update your bathroom floor with ceramic tile (page 70).

Because items such as garage door openers and dishwashers wear out, we show you how to replace them, as well as other household items.

III. Woodworking Projects. We know that you are always looking for projects and project ideas. This year's selection of woodworking projects will knock your socks off! We include everything from a bathroom organizer (page 98) to a backyard gazebo (page 100). You'll find more than 17 projects from which to choose in this section.

IV. Tools & Techniques. Home improvement and woodworking requires tools, tool know-how and construction techniques. This section aids you in selecting the right sandpaper (page 164), buying a circular saw (page 174) and operating a thickness planer (page 167), to name a few.

This year we are particularly proud of the story on welding for the do-it-yourselfer (page 152) which offers instruction on how to use an inexpensive wire welder. Wire welding is easy to learn and very useful for a variety of home and automotive projects.

V. Home & Auto Management. Last, but not least, this section helps you manage your home and automobile better. Something as simple as replacing a thermostat can save precious fuel and in the process save money.

You'll see a difference in this year's *Yearbook* — a new look. We have gone to a three-column format for easier reading and to enhance the page's look. Unlike previous years, the sections are better separated with divisional pages. We are currently planning next year's *Yearbook* to look and be even more functional.

We appreciate our readers and are always interested in hearing from you. Please direct your comments to Popular Science® Do-It-Yourself, P.O. Box 19, Cannon Falls, MN 55009.

Thank you for your dedication to our *Yearbook*.

Al Gutierrez

Contents

Six Homes
for Comfortable Family Living

Looking at home designs stirs our imaginations, and compels us to dream of our ideal home. It's even better when we can capture those design images in pictures. For your enjoyment and stimulation, we are pleased to present these six home designs. The six plans are available from a plans supplier (see address at the end of the story). — *by Al Gutierrez.*

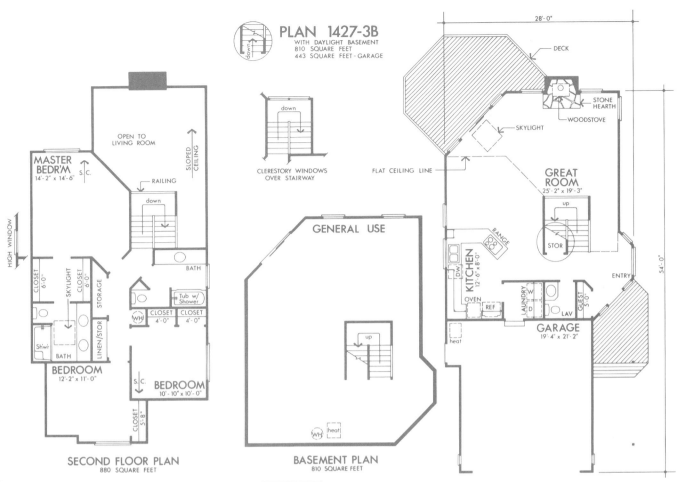

PLAN 1427-3B
WITH DAYLIGHT BASEMENT
810 SQUARE FEET
443 SQUARE FEET - GARAGE

CLERESTORY WINDOWS
OVER STAIRWAY

SECOND FLOOR PLAN
880 SQUARE FEET

OPEN TO
LIVING ROOM

SLOPED CEILING

MASTER BEDR'M
14'-2" x 14'-6"

S.C.

RAILING

down

BATH

HIGH WINDOW

CLOSET 6'-0"

SKYLIGHT

CLOSET 6'-0"

STORAGE

LINEN/STOR.

WH

CLOSET 4'-0"

CLOSET 4'-0"

Tub w/ Shower

Sh'wr

BATH

BEDROOM
12'-2" x 11'-0"

S.C.

BEDROOM
10'-10" x 10'-0"

CLOSET 5'-8"

BASEMENT PLAN
810 SQUARE FEET

GENERAL USE

up

WH

heat

PLAN 1427-3A
WITHOUT BASEMENT
810 SQUARE FEET
443 SQUARE FEET - GARAGE

28'-0"

DECK

STONE HEARTH

WOODSTOVE

SKYLIGHT

FLAT CEILING LINE

GREAT ROOM
25'-2" x 19'-3"

up

STOR

RANGE

KITCHEN
12'-6" x 8'-0"

DW

OVEN

REF

ENTRY

LAUNDRY

W
D

LAV

GUEST 5'-0"

54'-0"

GARAGE
19'-4" x 21'-2"

heat

M^cLURE

Ideal Home for a Narrow Lot

Because this design is only 28 ft. wide, it could be built on a 40 ft. lot, local codes permitting. It provides 1,690 sq. ft. of living space.

An entry hall provides the introduction to the wide-open arrangement of the first floor. One is immediately drawn to the balconied staircase and bay window near the entrance.

On the first level, the great room, dining area and kitchen combine so that no one need be excluded from activities or conversation. Yet each area has its separate identity: the great room with its fireplace, the dining area with its large outdoor deck visible through sliding glass doors, and the U-shaped kitchen separated from the other areas by the breakfast bar.

Upstairs are three bedrooms and two baths. The master bedroom suite includes a large walk-through closet in addition to a private bath.

3

BASEMENT PLAN
605 SQUARE FEET

PLAN 930-1A
WITHOUT BASEMENT
1210 SQUARE FEET

PLAN 930-1
WITH BASEMENT
1210 SQUARE FEET

Prowed Chalet

This design offers versatile living space that can be expanded as family needs change. The second-floor balcony area can be a sitting room or a fourth bedroom. The downstairs bedroom can be used as a TV room or a den. The house is suitable to level or sloping lots. What's more, the design allows the owner-builder to work in stages, finishing the second floor or the basement as time and budget allow.

The home is characterized by the prow-shaped living room and surrounding deck. Sliding glass doors in the living room and a side entry make the deck accessible from all of the home's main living areas. The 1,210 sq. ft. first floor is enhanced by vaulted ceilings, a second-floor balcony, and skylights.

The kitchen is open to the living room. Another outside entrance at the end of the central hallway provides easy access to the full bath and the laundry room, which can also serve as a mudroom.

The second floor plan of 710 sq. ft. includes two bedrooms and a balcony room that overlooks the living room. A second full bath is also located on this level.

SECOND FLOOR PLAN
710 SQUARE FEET

Compact and Luxurious

This home bundles up into one compact package all the best from the past and the present. The shed roof is reminiscent of a New England saltbox, while the gabled dormers, wooden louver vents and half-circle windows recall the Victorian era.

Beyond the charming facade, its 2 x 6 construction allows R-19 insulation in the walls.

The cozy kitchen has a center island with a breakfast counter and built-in range/oven. The corner sink, framed by windows, saves on counter space. A built-in pantry and a china closet are centrally located between the dining room and the kitchen. The nook with bay window is popular for everyday meals. The formal dining room, separated from the living room by a railing, affords a view to the living room and fireplace beyond.

The large living room has a vaulted ceiling and built-in shelves for an entertainment center. Sliding glass doors on one side of the fireplace lead to the backyard deck. To the left of the fireplace is a picture window topped with a half-round window.

Sliding glass doors in the larger of the two bedrooms provide additional access to the deck.

The full bath is handy to both the main living area and the two bedrooms.

The second floor is devoted to the master bedroom suite, which features a hydro-spa, shower, vanity and walk-in closet. The stairway is open to the living room below.

This home has a total of 1,771 sq. ft. of living space.

MASTER BEDROOM
13/0 x 15/0

OPEN TO LIVING ROOM

down

9/6

WALK-IN CLOSET 4/0

BATH

HYDRO SPA

Shr

SECOND FLOOR PLAN
386 SQUARE FEET

PLAN 1453 –1A
WITHOUT BASEMENT

58'-0"

DECK

FIRST FLOOR PLAN
1385 SQUARE FEET
438 SQUARE FEET (GARAGE)

LIVING ROOM
24/6 x 14/0
SLOPED
CEILING

BEDROOM
13/0 x 13/0

41'-7"

GARAGE
19/0 x 21/6

SHELVES

W.H. furn

UTILITY

D

W

PAN

DINING
11/6 x 10/0

down

up

CLOSET
9/6

LIN

tub w/sh

BATH

CLOSET
7/9

ENTRY

NOOK
7/6 x 11/0

R/O

GLASS

KITCHEN
10/0 x 12/0

P.W.

REF

BEDROOM
11/0 x 11/0

up

Build It on Weekends

PLAN 921-1A
1164 SQUARE FEET

42'-0"   8'-0"

30'-0"

BEDROOM
11'-6" x 13'-6"

BATH

w d

LAUNDRY

DINING

KITCHEN
9'-7" x 8'-2"

LIVING RM
12'-0" x 23'-0"

heat

BATH

wh

CLOSET

CLOS CLOS

LINEN

ENTRY

BEDROOM
11'-6" x 10'-3"

BEDROOM
10'-9" x 10'-0"

CLOSET

CLOS

DECK

This 30 ft. by 42 ft. rectangular dwelling has an 8 ft. deck at one end that wraps around one side of the living room and entry hall. The simple shape of this building enables the use of truss framing in the roof, for ease and speed of erection.

The level ceiling throughout allows the builder to install as much as 12 in. of insulation in the ceiling; 2 x 6 studs in the perimeter walls permit the use of R-19 insulation. The inside surface of the foundation walls is R-11. Heat transfer is further inhibited by double-pane insulating glass used throughout the home.

One of the most interesting features of this design is the use of a crawlspace plenum as an air distribution chamber. Referred to in the construction industry as a Plen-Wood system, it utilizes the sealed crawlspace as a chamber for the distribution of heated or cooled air, which is forced through floor ducts by a downdraft furnace. This gives the home a warm, comfortable floor.

Tests have shown that the Plen-Wood system produces more uniform air temperature, floor to ceiling, than other conventional heat duct systems. Therefore, the thermostat can be set lower in winter and higher in summer.

Other features that make this home a desirable construction project include clustering the two bathrooms, laundry room and kitchen, which allows common plumbing walls. An excellent traffic plan minimizes cross-room traffic and hall space. Built-in storage closets and wardrobe space are abundant.

The furnace is located at the center of the home, making it an efficient source of heat.

A see-through kitchen has open counter space so the view to the living room is unobstructed. Sliding glass doors are installed on the long outside wall of the living room, providing easy access to the covered deck. A prefab metal core fireplace is built into the end wall of the living room, and sheathed in wood that matches the exterior siding of the home. Living area of this dwelling is 1,164 sq. feet.

Hillside Design
Fits Contours

This split-level design takes maximum advantage of the natural contours of a building site. Develop grade-level entrances and outdoor living areas with this plan, a big plus in these days of indoor-outdoor living. It has 1,533 sq. ft. of living area.

This plan also provides excellent separation of living areas from sleeping quarters. The master bedroom with walk-in closet measures 333 sq. feet. The personal bath includes a stall shower and built-in vanity. A sun deck is accessible through sliding glass doors.

Two more bedrooms are on the first floor, along with a full bath. Bedrooms are separated by the spacious entry hall that connects with the living room, dining area and U-shaped kitchen. A wrap-around deck is built across the rear of the home. A central hall provides direct routes to every area.

Practical kitchen and laundry areas have outside access. A staircase leads to the lower daylight basement that includes a fourth bedroom with walk-in closet and complete bath. The basement also includes a spacious recreation room with doors connecting to the rear garden level. An abundance of basement storage is provided by a room that spans the 39 ft. width of the house and is 13 ft. in depth.

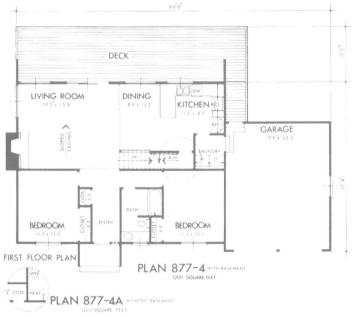

DECK

LIVING ROOM
19'3" × 15'6"

DINING
8'9" × 12'3"

KITCHEN
11'0" × 9'0"

GARAGE
19'9" × 23'3"

SLOPED CEILING

LAUNDRY
W   D

BEDROOM
11'3" × 13'0"

CLOS
3'0"

CLOSET
6'6"

BATH

ENTRY

CLOSET
4'0"

BEDROOM
11'3" × 13'0"

FIRST FLOOR PLAN

WH

STOR

HEAT

PLAN 877-4 with basement
1200 SQUARE FEET

PLAN 877-4A without basement
1200 SQUARE FEET

DECK

BEDROOM
13'9" × 12'3"

WALK-IN CLOSET
5'3" × 4'9"

LIVING ROOM BELOW

Shwr

BATH

down

SECOND FLOOR PLAN
333 SQUARE FEET

PATIO

BEDROOM
13'9" × 12'3"

WALK-IN CLOSET
5' WIDE

RECREATION
19'11" × 15'6"

STORAGE

Shwr

BATH

GENERAL USE
38'9" × 12'9"

PLAN 877-4B with daylight basement
1200 SQUARE FEET
BASEMENT PLAN

9

Where to Order Plans

To order complete construction blueprints for any of these six house designs, contact HomeStyles Plan Service, 275 Market St., Suite 521, Minneapolis, MN 55405, (800) 547-5570.

All-American Country Home

This plan features a wrap-around porch that is as functional as it is decorative. The covered porch is 8 ft. wide, providing plenty of room for outdoor dining, entertaining and relaxation. It extends from the front door to the French doors in the rear, which open onto the living and dining area. The post-and-rail detailing softens the home's rectangular shape and gives it a country flavor.

This country home has 2,464 sq. ft. of living space.

Inside, that old-fashioned charm blends with modern-day convenience. The large country kitchen features an island counter for serving and preparing food. Cabinets and appliances are grouped around the island in a step-saving U-shape. A natural gathering spot, the kitchen is open to a large family room for additional seating. This comfortable room is highlighted by exposed beams and a raised-hearth fireplace.

The more formal living and dining rooms are separated from the informal kitchen and family room by a large foyer and the stairwell. Another fireplace graces the living room, while the dining room has picture windows overlooking the backyard.

At the opposite end of the house, between the kitchen and the garage, is the laundry room. This fully-equipped room includes cabinets over the washer and dryer, and a utility sink. An entry door leads from the laundry room to the garage and storage area.

All four bedrooms are on the second floor, completely separated from the living areas. The master suite features a walk-in closet, dressing room and private bath.

This design is available with or without basement, and with or without garage. The basement versions add another 1,176 sq. ft., including a recreation room with fireplace. All versions feature 2 x 6 exterior wall construction.

SECOND FLOOR PLAN
1176 SQUARE FEET

PLAN 3711-1 WITH BASEMENT
1288 SQUARE FEET
PLAN 3711-1A WITHOUT BASEMENT

Steel-Framed Construction for Houses

Steel-framed construction is rapidly making its way into the home-building business. Here's why.

If you've ever played "sidewalk superintendent" while a new commercial building has gone up, or if your company has remodeled its office space, you are aware of the increasing use of steel in framing products. Steel studs and partition members are lightweight, easily assembled (using the proper tools), and straight and warp-free. They also meet strict fire codes. The 9 in. depth of the steel wall studs makes it possible to insulate side walls with up to R-30 insulation. Plus, new fasteners such as self-tapping screws driven by power screw guns make it possible to attach any finish material to the steel framing.

With all these advantages, it is no wonder that steel-framed construction is increasingly being used in residential buildings. Yet many people are wary of steel-framed homes, believing that they are not as sturdy or as attractive as wood-framed houses. Nothing could be further from the truth. Steel-framed houses don't look "different" any more; in fact, that attractive new home in your neighborhood just might be framed with steel. What's more, the strength and energy-efficiency of steel-framed construction rates at the top of the list. Consumers are already well aware of the steel sandwich doors with foam insulation cores that increase energy efficiency. Steel doors and frames also enable you to use magnetic weather stripping that creates a thermal seal just like the one on your refrigerator door.

Add to the list of benefits the fact that erection crews can frame a steel house quickly. Bolting together the fabricated framing members means you can get the building "under roof" or enclosed against the weather quickly, a great advantage when building in cold or wet

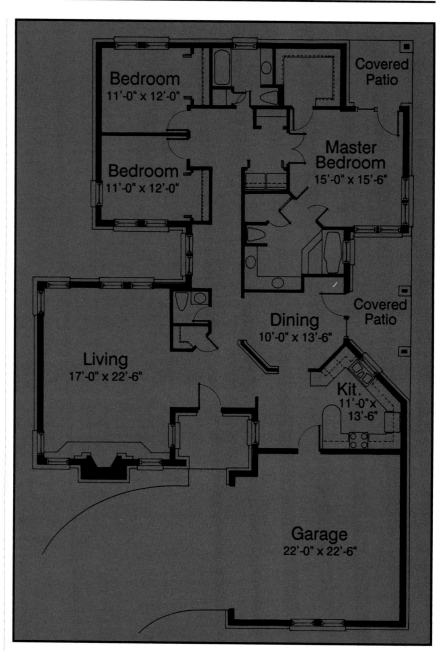

1 *Like any house project, the first step is to pour the slab or build a foundation or basement. Allow concrete to cure.*

2 *The construction crew bolts the steel-framing components together. This step goes quickly.*

3 *Steel frames are laid out on the ground for tilt-up erection.*

4 *Anchor bolts set in concrete around the perimeter of the house's foundation hold the framing in place.*

5 *A crane lifts the steel framing into position.*

weather. You can then proceed with interior finishing at your leisure, in a dry, climate- and temperature-controlled environment. The dry climate plus the steel framing prevents both nails and fasteners from loosening or "popping" and eliminates problems with rusty fasteners.

Another plus for steel-framed homes is their exceptional strength. Steel-framed houses have been shown to resist damage from 100-mph wind loads, which definitely recommends them for use in areas where hurricanes or tornadoes are common. Steel framing has also demonstrated superior strength for resistance to earthquakes, up to 4 or more on the Richter scale. Add to these advantages the natural resistance to fire and termite damage and you may find yourself taking a close, hard look at steel framing for your next home.

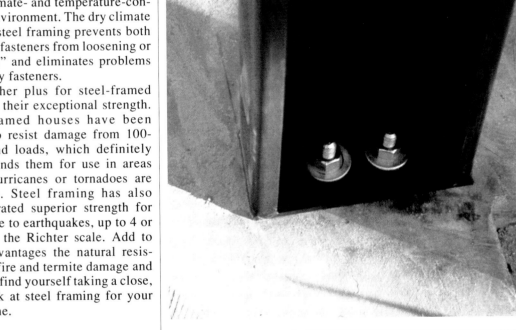

6 *Frames are set over the foundation bolts and secured with nuts.*

7 *Purlins (cross braces) are attached to hold the frames.*

8 *Self-tapping (drilling) fasteners are used to attach the purlins to the frames.*

9 *Intermediate steel studs are installed between the frames.*

10 *Plywood sheathing is screw-attached to the frames.*

11 *Fiberglass batt insulation is installed between the framing members.*

12 *Windows and doors are installed per the manufacturer's specifications.*

13 *Exterior shell is complete and weather-tight.*

14 *The construction crew or owner can proceed with interior finish chores. Superior results can be achieved in this dry, climate-controlled environment.*

15 *The home's exterior finish can be brick, stucco or any type of siding.*

16 *The last step is exterior grading and landscaping. The owner can finish these tasks as time and money permit.*

Companies such as Tri-Steel Structures, Inc. of Dallas, Texas, have been refining the steel construction process for almost 15 years. Their research and development of steel framing in the residential marketplace has yielded a "builder friendly" product that is economically viable, and has met the test of time and with the approval of building officials in all parts of the country. Steel provides the straight walls, level floors, flat ceilings and square corners that are the hallmarks of the quality builder.

Economical, efficient, sturdy and attractive. What more could anyone want for his next home? For more information about steel framing in residential construction, call 1-800-TRI-STEEL. — *by Gary Branson. Photography by Tri-Steel.*

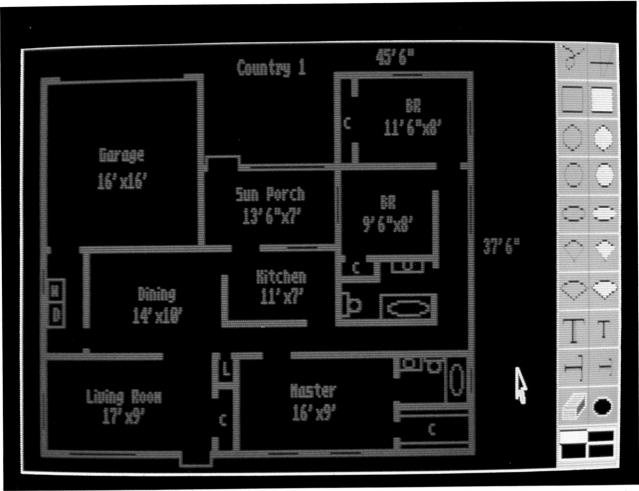

Architecture's screen is shown here with one of the supplied floor plans. You can design everything shown here. A clip art menu is supplied to help you illustrate doors, bathtubs, showers, etc. You can even select the colors you desire.

Design Your Own Home

Here's how you can make use of your personal computer to design your next home.

Did you buy a computer with the intention that it would expand your knowledge and efficiency only to find that it sits around? Maybe it's not the computer that you find uninteresting, but the programs you have to use for it. If that's the case, you'll be happy to learn of a new program called *Design Your Own Home: Architecture*.

Designed for an IBM or IBM compatible computer, this fun program lets you create floor plans and house elevations in very little time. It can be used for designing a new home or for expanding an existing home. By drawing a series of squares and rectangles you can generate the areas making up each room. Then add doors, windows, showers, sinks,

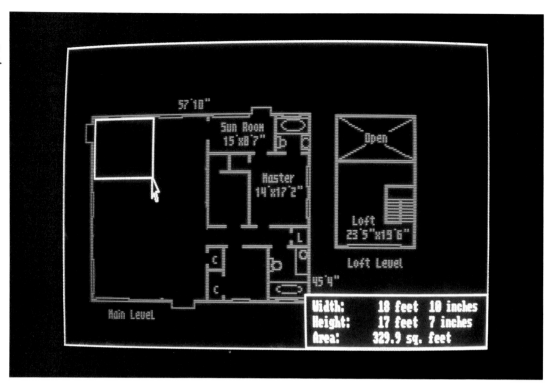

etc., by selecting them from a clip art menu. Complete the drawing by filling in rooms or landscape with colors.

You will need an IBM or IBM compatible computer equipped with 640K of RAM, a 2.1 or higher version of DOS, a graphics card and a mouse. Print out your results to a dot-matrix or laser printer. Of course, a color printer is ideal because it allows you to output your selected colors. It works for computers with or without hard drives.

Architecture comes with more than two dozen sample floor plans and elevations, some of which are illustrated here. If you don't like the sample floor plans and prefer to create your own plans, *Architecture* gives you that option. You can modify the sample floor plans simply by erasing unwanted walls, windows, etc. The program also comes with an easy-to-read 44-page owner's manual. Most of the information is non-technical and geared to the novice.

By the time you're finished reading the owner's manual you should be creating basic floor plans with the aid of a mouse. Dimensional information appears on screen to help you draw the proper room size. Along with this dimensional information is a calculation of the room's square footage. You will no doubt have to refer to the owner's manual quite often, but you should catch on very quickly.

This program is not intended to replace professional architectural drawings. Though the floor plan and elevations are very useful, architectural dimensioned details, like wall sections, are still needed for construction.

Architecture sells for about $100 and is also available for the Apple and Macintosh computers. For more information or to obtain the program, contact Abracadata, P.O. Box 2440, Eugene, OR 97402, (503) 342-3030. — *by Al Gutierrez.*

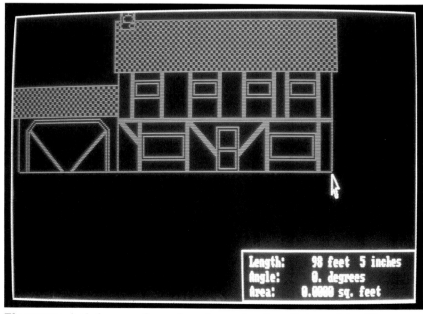

The program includes more than a dozen house elevations that correlate with supplied floor plans. These may be modified to suit your needs.

The All Masonry Foundation

Enhance your next home's basement with architectural block.

Planning your first home, or your next one? Consider enhancing the basement with masonry. Choosing concrete masonry gives you a world of options. Architecturally-faced concrete masonry, also known as architectural block, is now available in a variety of styles that give you limitless design options. The concept is simple to appreciate: the finished product looks great, but does not add greatly to the cost of a home. Let's examine how concrete masonry basements offer a consumer more "bang for the buck."

First, it is important to note that below-grade floor space is inherently less expensive than the same amount of floor space above grade. The message to consumers is that they can hardly afford to pass up the chance to add functional space to their homes at attractive prices.

Basements need not be dark or dull or colorless. A wide range of colors in common and architectural block have been a standard on the American landscape. Office buildings, hospitals and apartment complexes have used architectural block inside and out for years. And these are just as attractive, and offer just as much appeal, in basements.

You can give a basement the look of a European castle, or use a series of vertical lines to break up the space in a long horizontal wall. It is easy to combine various architectural block styles into one wall because the sizes are consistent and they are installed by hand. Concrete masonry is so versatile that over

Cost of Basement Impact Analysis
(Assume 1,200 sq. ft. for 1 story, 1,800 sq. ft. for 2 story)

Residence Type	Base S.F. Cost	Add Finished Basement	Increase in S.F.
1 Story econ.	$48.00	$ 9.00	100%
2 Story econ.	42.00	6.00	50%
1 Story average	55.00	12.00	100%
2 Story average	53.00	8.00	50%
1 Story custom	75.00	19.00	100%
2 Story custom	68.00	12.00	50%

1,400 combinations of shapes, sizes and colors are attainable.

Not only do walls of architectural block look great, but they also have all the watertight, energy-efficient and fire-resistant advantages of standard block walls. The added attraction with architectural block is the commanding presence of beautiful lines, sensational textures and warm colors.

The architectural block basement causes no construction headaches for the builder. It has all the advantages of building with block, and adds an attractive finish to the below-grade space for only pennies a square foot. There's no need for drywall hangers or painters. Once the wall is up, it is finished.

Looking at the Numbers

Basements are a good deal on almost every homeowner's bottom line. The homeowner pays for the slab, whether it is on grade or below grade, so that cost doesn't change. In many cases, a crawl space is included, so the only additional requirement to complete a full basement is deeper excavation, and five to eight extra courses of block. The added expenses are moderate.

Let's take a look at the actual building costs, to the builder, to demonstrate what the consumer pays for. Take a typical 1,200 sq. ft. rambler built for the starter market. According to Mean's Residential Cost Guide, construction expense is going $48 per sq. ft. for the basic house with no frills, or $57,600 to the builder. For $61,500, the builder can erect the same house with a 1,200 sq. ft. concrete masonry basement, to bring the cost per sq. ft. down to $25.62. The basement doubles the size of the house at an additional cost of only $3,900.

In general, no matter what the size of a one-story home, by spending approximately one-fifth the above-grade construction costs of an average-quality single-story home, you gain 100 percent added space by building a basement. Spend 15 percent more on a two-story home, and the space is increased by 50 percent.

Why Block Basements?

Concrete masonry is a product of simple virtues. It is strong, durable, comes in a variety of colors and shapes, and is easy to work with. Most builders are aware of one of its more attractive features — low cost of the installed wall. The average price of a 12 in. concrete masonry wall built with standard procedures is $6.63 per sq. foot. A 12 in. wall of poured concrete would cost $7.18 per sq. foot. With a $0.55 per square ft. difference, the cost is reduced by

Basement Wall Comparison Costs Per Square Foot
(Includes labor but not contractor overhead or profit.)

12 IN. BLOCKWALL

Concrete masonry unit	$4.50
Wire reinforcing	.30
Parging	.45
Damp proofing (1 coat)	.35
1 in. board insulation	.65
Anchor bolts	.08
2 x 4 sill plate	.30
Total	**$6.63 per sq. ft.**

12 IN. POURED CONCRETE

Concrete	$2.20
Formwork	$2.80
Light reinforcing	.40
Concrete placement (chute)	.40
Damp proofing (1 coat)	.35
1 in. board insulation	.65
Anchor bolts	.08
2 x 4 sill plate	.30
Total	**$7.18 per sq. ft.**

$660 on a 1,200 sq. ft. basement.

A major reason that a block wall costs less than a poured concrete wall is that the only subcontractors needed are masons. Once the site is excavated, two mason crews can build a basement in two or three days. Pouring a basement requires one day for the form crew and a crew to pour the concrete, which must be left to cure for several days before construction can continue.

Block also allows a choice of a wide variety of insulation techniques: preinsulated concrete masonry units, loose-fill granular insulation on site or installing rigid insulation boards along the outside wall. Insulated concrete masonry basement walls typically have an R-value between R-10 and R-18. Block basements are also more crack-resistant and are easily waterproofed.

Concrete masonry is adaptable to any house design, thanks to its small unit size, manual installation and the wide range of special shapes and sizes available. This design flexibility allows the builder or contractor to offer a wider range of home styles without increased costs and construction time.

Construction Techniques

The architect may take advantage of the modular nature of concrete masonry by designing the floor plan of the house as close as possible, using the linear dimensions of the block. Of course, it won't always be possible to maintain the modular proportion — 15⅝ in. for the block plus a ⅜ in. mortar joint, or a total of 16 inches. But the closer to this ideal, the better, because it reduces the amount of cutting and fitting the mason must perform on the job. A basement wall will still be strong and watertight if blocks are cut, but cutting and fitting slow down the job, and therefore make it more expensive.

From a practical point of view, it is impossible to predict the amount of moisture that a foundation will be exposed to. But if water is controlled at its source, there will be no trouble with leaking. To relieve hydrostatic pressure, for example, install a subsurface drainage system. Placing drain tile around the perimeter of the foundation will do the trick.

To divert water and dampness from the exterior of the wall, apply a waterproofer and water barrier. And, to minimize accumulation of surface ground water, grade away from the basement on all sides.

Availability

Architectural block is widely used by some of the nation's biggest developers. The supply is plentiful and distribution is rarely a problem. Of the 900-plus concrete masonry manufacturing plants in the United States, better than half produce some kind of architectural unit, according to a recent National Concrete Masonry Association (NCMA) survey. — *by Ray Lorenz.*

This story is courtesy of Builder magazine and the National Concrete Masonry Association.

For further information, or the name of the NCMA member nearest you, write Bryan Earl, Director of Marketing and Communications, P.O. Box 781, 2302 Horse Pen Road, Herndon, VA 22070, or call (703) 435-4900.

With the hole already dug, this homeowner carefully hauls a young tree to its permanent planting site. These pine trees grew about 13 in. in two summers.

Start Your Own
BACKYARD NURSERY...

And cut the high costs of landscaping.

Although this backyard nursery was started in a rural setting, one can be successfully started in a smaller urban yard.

The term *tree nursery* isn't meant solely for professional landscapers who plant and cultivate 50 acres or more of trees and shrubs. You can develop your own home nursery, whether you have a spacious three-acre tract or a suburban 1/4 acre or smaller yard. In fact, a pint-sized 10 x 10 ft. garden plot or patch of extra space is room enough to plant several seedling spruces or whippy ash, maple, or fruit tree saplings.

Why start your own backyard nursery? Basically, it's a money-saver. Buying three to five 3 ft. high balled and burlapped trees or shrubs from a commercial nursery is expensive, not to mention the time and trouble in hauling them home. But purchasing seedlings in batches of 25, 50 or more from local or mail order nurseries is considerably less costly.

Ask your local nurseryman or extension service what trees are best suited for your region. Don't plant a southern magnolia in northern Minnesota, or a blueberry bush in alkaline soil. Each species requires a specific habitat.

Stay away from digging up the wild sapling you see along the road.

This home nursery was started from potted 8 in. pines during the fall of 1984 and were transplanted in the fall of 1986. Over the course of two growing seasons, the value of 24 pine trees grew from an initial cost of $96.00 to $360.00. The trees' value increased 375% on the average.

PLANTING SEEDLINGS

1 *When laying out your nursery, you should plant grass between the trees. The grass will curb erosion while giving you a mud-free path to walk on.*

2 *A roomy hole allows young roots to expand. Remember the old adage: "Don't plant a $5 tree in a 5 cent hole." The hole should be about twice the depth and diameter of the root ball.*

3 *Add 3 inches of loose soil to the bottom of the hole, then mix in a few handfuls of organic matter. The root hairs grow into the loose soil; the compost nourishes them. Now insert the tree.*

4 *Note how the trunk is planted at the same depth as it was in the peat pot. Also, leave a shallow trench (shallower than shown here) to catch and direct rainfall to the roots.*

5 *Water thoroughly every day for at least a month after planting.*

TRANSPLANTING TREES

1 *Spade deeply around the tree's outer circumference (drip line), and no closer. You'll lose a few root hairs, but try to keep loss to a minimum. Use a long-billed spade.*

2 *Gently push down on the spade's handle to loosen the root ball. Lift the root ball from underneath with your hands. Don't pull on the trunk because you might damage the roots.*

3 *Ease the tree into the center of the hole, flood it halfway, then mash with hands to remove air pockets in the root area. Add remaining fill and gently tamp it down.*

4 *Build a cone-shaped mound around the trunk to serve as a catch-basin for water. Always keep the base of the trunk free of grass and weeds. To prevent the tree from drying out, wrap the trunk with aluminum foil or strips of burlap.*

In most states, it's illegal to dig them.

Remember, your home nursery is an investment in the future. And the initial planting of small whips and seedlings is a temporary one. Within three to five years, the seedlings should be ready to transplant to their permanent spot. Meanwhile, you not only have a chance to blueprint where they'll be planted, but you'll have the sheer joy of watching them grow.

Plan Your Nursery

Make sure your nursery area sees at least eight hours of full sun per day, has good soil texture, fertility, and proper soil pH. Experts underline the importance of the latter three.

"Drainage and soil pH are the most vital, but the most often over-looked," says Edward Whiteman, a horticulturist with the Pennsylvania Nurseryman's Association, Harrisburg, Pennsylvania. "Spruces and pines won't tolerate constantly wet soil, so avoid planting in low spots."

As for soil pH and nutrients, Whiteman emphasizes the all-important soil test. "Soil test kits are available at most garden centers, or send a sample to your county agricultural extension service for a full report to see whether your soil is too acidic or alkaline, and what nutrients may be absent."

Whiteman adds, "Your soil should contain phosphorus and potassium for good root growth, and micronutrients, such as iron and magnesium, are essential for peak health. Pin oaks, a common species, are notorious for having an iron deficiency."

Whatever your soil nutrient and pH levels are, you won't have healthy trees and shrubs if the soil isn't friable (easily crumbled) and well-aerated. If you have farmland loam, you're in clover! If your ground is hardpan or mucky clay, rent a roto-tiller and load up the soil with organic matter.

Almost anything will build drainage and aeration — chopped straw or cornstalks, lawn clippings, shredded leaves, and peat moss. Don't scrimp. Pile it on and dig it in. Organic matter builds soil composition and adds required nutrients.

If you're tilling during spring, however, there is one caveat. "Horse, cow, chicken or sheep manure is ideal," advises Harold Pellette, a soil expert at the University of Minnesota Landscape Arboretum near Chaska, Minnesota. "But it should be well-rotted. Fresh manure ties up too much nitrogen and makes it unavailable for root and leaf growth."

Pellette and other experts say autumn is best for soil tilling because the manure and other organics will decompose over the winter. However, spring tilling is fine as long as the organic matter is dried and mostly decomposed.

Proper Planting Methods

With the soil in good shape, you're ready to plant. There's nothing mysterious or highly scientific about planting a tree — just common sense and patience.

First, the old adage, "Don't plant a $5 tree in a 5 cent hole," is true. The hole should be twice the diameter and depth of the seedling's root ball (small seedlings usually are rooted in peat pots, but occasionally are bare-rooted). Loosen the soil at

the bottom of the hole to about 3 inches. This allows the roots to establish good contact.

Next, center the root in the hole, toss in a few handfuls of organic material, then slit the pot with a knife. Bury the peat pot with the roots when refilling the hole. When the hole is half-filled, flood the hole with water, let it settle, then work the soil/water mix with your hands to eliminate air bubbles and pockets.

Bury the tree so it is at the same depth as it was in the planter pot, and completely refill the hole with soil. Edward Whiteman of the Pennsylvania Nurseryman's Association cautions: "Planting the tree an inch too high is better than an inch too low. If the trunk is too low, disease can enter the tender bark."

The University of Minnesota's Harold Pellette advises thorough and regular watering after planting. "It's essential for solid root growth, but if your soil remains heavy, avoid watering more than once a week. That's about all the water a tree will absorb and remain vigorous."

Forget the Fertilizer

One other precaution: Don't fertilize your trees for the first year that you transplant them. Chris Price, a horticulturist at Bachman's, Inc., a well-known nursery/landscaping firm based in Minneapolis, Minnesota, puts it this way: "A newly planted tree must grow root structure. A dose of chemical fertilizer urges the tree to produce foliage and stem growth, but the roots won't bother to spread," he says. "Why should they? They have plenty of fertilizer to feed on. Extra fertilizer merely creates a bonsai tree — all foliage and small roots."

Most experts recommend pruning any spring-planted maple or other deciduous tree by one-third. Most any deciduous or coniferous tree can be pruned if it's 2 ft. or higher. Also wrap the trunk to protect it from the sun.

Extra Care

Once your nursery is established, just follow a routine of maintenance. Keep the base of each tree mulched, and water well. Experts strongly

advise keeping grass from encircling the trunks, since it can limit tree growth.

"Turf shuts out moisture and nutrients," states Ed Whiteman, "and it slows plant growth phenomenally. Homeowners wonder why their trees see scant growth, and encroaching turf is a major reason why."

According to Whiteman, trees can have grass between each row. This slows erosion and gives you a mud-free place to walk. But leave a 2 ft. wide grass-free circle around the trunk.

How to Transplant Your Trees

After three to five years, well-started trees can be dug and transplanted to a permanent location. At this point, digging out the now 4 to 5 ft. saplings requires a bit of finesse.

Take a long-billed spade and dig around the drip line of the tree. You'll lose a few root hairs, but you should keep the loss to a minimum. Pro nurserymen dig a trench around the tree, then water the trench thoroughly. This method keeps the root ball solidly together. Dry roots might break or lose too much soil when lifted.

Next, slip a sheet or tarp under the root ball and carry the tree hammock style to the new planting site. It's a simple procedure, but don't rock the tree back and forth, or the trunk may break away from the roots. At the new site, gently slide the ball into the new hole. Plant it the same way as you did when the tree was a whippy seedling.

Keep it watered and mulched, and stake the trunk. A transplanted sapling is much larger than it was during its seedling days, and its thicker foliage catches the wind easily. A staked tree is less prone to damage from strong winds.

With the right care, a healthy tree will provide shade for future generations. Better yet, right now it will add beauty and value to your landscape. — *by Ray Lorenz.*

Sources:
Garden tiller, shovels, wheelbarrow: Sears Catalog.

Replace a Wax Ring

Learn when and how to replace your toilet's wax ring.

Have a leak around your toilet? It could be the wax ring that seals the connection between the bottom of the toilet and the drainage pipe.

Water leakage around the base of a toilet, especially around the front, could be indicative of a bad wax ring.

To determine the cause, first check for any cracks in the stool.

Also check the water connections entering the base of the tank.

If the leak gets worse when you flush, then it is probably due to a cracked toilet or a bad wax ring. To further diagnose the problem, wipe the area around the toilet dry, then flush the toilet and check for leaks. If water continues to puddle at the base of the toilet but the stool itself is dry, then replace the wax ring.

To replace the wax ring you have to shut off the water supply, remove the water tank and then remove the toilet. Follow the step-by-step photos for replacing the wax ring.

You can buy wax ring replacements at hardware stores and home centers. Also buy plumber's putty and a new gasket to seal the water tank to the toilet. — *by Gene and Katie Hamilto*n.

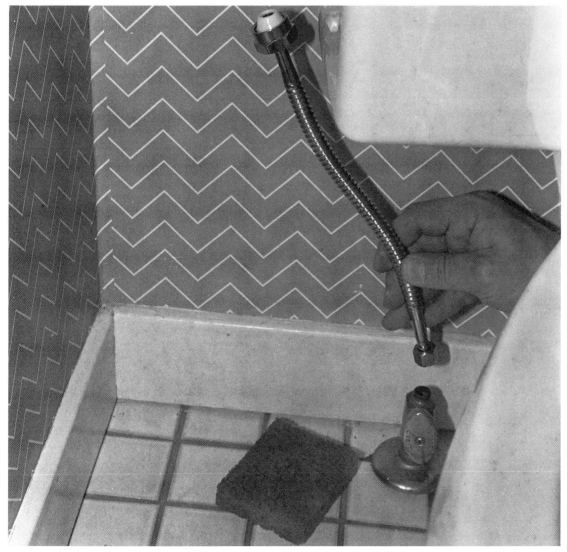

1 *Turn off the water supply and disconnect the supply line that connects directly to the water tank.*

2 *Plunge the toilet to force the water down the drainage pipe. Then use a sponge to remove the remaining water. If you do not do this, the water will leak all over the floor when you remove the toilet.*

3 *Remove the water tank by loosening and removing the bolts connecting it to the stool. You will also have to prevent the bolt from turning on the inside of the water tank by backholding with a screwdriver or a wrench.*

4 *Lift the water tank straight up after you have removed the bolts. Set the water tank aside.*

5 *Scrape away all the putty around the base of the toilet. Use a putty knife and avoid damaging the floor.*

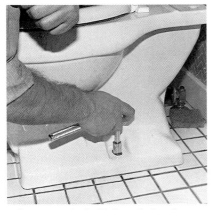

6 *Remove the hold-down bolts after cleaning the threads with a wire brush. You may have to apply several applications of penetrating oil to loosen frozen nuts.*

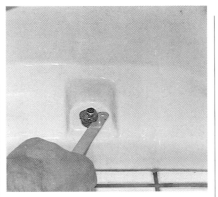

7 *If the hold-down bolts are still frozen, even after applying penetrating oil, you may have to cut the bolts with a hacksaw blade. Place the blade under the nut and work in short strokes.*

8 *Remove the seat first, since it is easier to maneuver than the toilet. Then carefully lift the toilet and relocate it where you can work on it. Tilt the toilet on its side. Refer to next photo.*

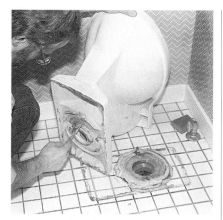

9 *Remove the old wax ring and scrape away the old wax and putty from the bottom (including edges) of the toilet and the drainage pipe. Clean the area thoroughly with a wire brush.*

10 *Place the new wax ring over the opening in the bottom of the toilet. Then twist the wax ring to help seat it.*

11 *Apply plumber's putty around the wax ring for additional sealing. Also, put putty on the outer edges of the bowl to provide a water seal to the floor.*

12 *Place some plumber's putty in the bolt slots of the drainage pipe to hold the bolts upright while you reposition the toilet.*

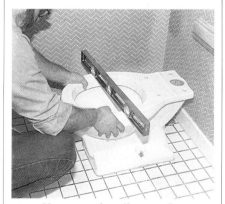

13 *Now move the toilet over the drainage pipe and bolts, and gently position the toilet. Push the toilet down and level it as shown.*

14 *Coat the water tank's bolts with Vaseline to prevent corrosion and replace the rubber seal.*

15 *Install the water tank and gently tighten the bolts. Make sure you have a good connection. Do not overtighten, or you could crack the tank.*

16 *Finally, after several days, retighten the toilet bolts and install the bolt covers. Place putty in the base of the covers to hold them in place if they do not snap into position.*

Install a Bay Window

An easy, low-cost solution to making your dining room brighter and more spacious.

This project is courtesy of Jonathan Press.
Illustration by Workbench magazine.
Photography by Andersen Corporation
and Knox Lumber.

Cramped, tight and small describes many dining rooms, but squeezing just a few more square feet of space to make the dining area more comfortable can easily cost you a bundle.

One unique, less costly solution to gaining 8 sq. ft. or more of useful dining space is adding a full-length bay window. This easy-to-install bay window project takes advantage of the bayed area that is projected beyond the wall. The result is a larger dining area and a room that is illuminated with light. Best of all, the project costs you only about $2,000 if you do it yourself.

This Andersen bay window (No. 45-3452-20) is 5 ft. 6¼ in. high by 7 ft. 1¼ in. wide by 1 ft. 8¹⁵⁄₁₆ in. deep. It includes extension jambs, seatboard and screens.

Make Rough Opening

Begin by carefully reviewing the installation instructions that accompany your bay window. Then mark the window's proper rough opening on your house's exterior wall. Use a long level to draw the vertical lines and a chalk line to snap the horizontal lines.

Important: Turn off the power to all electrical lines that hook to or pass through this wall by tripping the circuit breakers at the main service panel. Then install an old blade in your circular saw and set the depth-of-cut so it will cut through the siding and sheathing to the stud wall. Wear good eye goggles, and make a test cut. Readjust the blade (if necessary), and cut the rough opening. Remove the siding and sheathing with a wrecking bar. Also remove the insulation at this time.

Remove the siding and sheathing below the rough opening as well. You will have to work carefully so you do not cut into the concrete or masonry foundation.

Determine where the window's header is to be located, and cut off the tops and bottoms of all studs with a reciprocating saw. You will have to cut into the soffit in order to access the upper wall and to insert overhead

This bay window (photo, far left) fulfills three needs: It expands the dining area by 8 sq. ft., makes the room brighter and provides a 180 degree vista of the countryside. From the outside (left) this do-it-yourself installed bay window blends in nicely with the rest of the house and requires only one to two days to install.

31

insulation later. If there is an existing window, remove all interior trim. Then use a wrecking bar to remove the studs along with the existing window. (Try to keep the inside drywall intact.) The drywall will help to keep airborne dust to a minimum. Cut the sole plate with a reciprocating saw and remove.

Later, you will need to insert the plastic side flange (supplied with the window unit), which will snap into the window and fit under the siding. Therefore, you must remove an additional 1 in. of siding (only) from both the left and right sides of the opening. Cut this siding back from the rough opening with a circular saw, but leave the sheathing intact. Try to fit the side flange under the siding, removing any nails that get in the way. Next, cut and install the header along with the supporting studs.

Window and Floor Prep

Remove the bottom sheet of plywood that comes with the bay window, and use it for the subflooring. If the plywood's ¾ in. thickness does not match the present floor's, then mark and cut the plywood to mount flush to the present floor.

At this point you have two options for expanding the floor to support the window unit. You can add a prefabricated floor joist assembly, or you can install shortened extension joists nailed to the present joists. If you install shortened joists, you must remove the outside header joist, and securely nail the shortened joists to the existing joists. We opted to prefabricate an assembly.

To prefabricate the floor joists, use the plywood panel you cut to suit the floor as a template for cutting the joists. Form the perimeter first, using 2 x 10s or 2 x 12s, and then install the joists on 16 in. centers. Make sure the assembly is well nailed. Next, nail the plywood to the assembly.

Pre-drill holes for attaching the assembly to the house with lag screws. If appropriate, make sure you pre-drill holes for running the electrical cable. Attach the joist assembly to the house, making sure the

floor is level. If it is not, you will have to insert shims where necessary.

Now, construct the rough sills, sole plate and cripple studs on which the window will rest. Notice how the rough sills overlap to provide a strong joint. Drill holes into the studs for electrical wiring.

At this point you may opt to install the window the next day. If so, cover the opening with plastic.

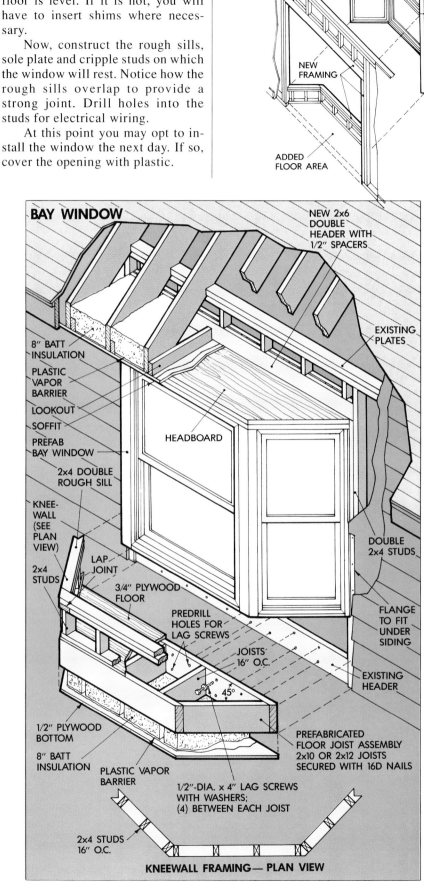

1 *Lay out the exterior's rough opening and cut out with a circular saw equipped with an old blade. Wear your eye goggles.*

2 *Cut the existing floor plate and studs with a reciprocating saw. Use a metal cutting blade for cutting through nails.*

3 *Cut and assemble the floor joist extension with 16d nails. An air driven nailer speeds assembly. Pre-drill mounting holes.*

4 *Install the joist assembly with lag screws. The floor should be flush with the existing floor's underlayment. Shim the unit as needed.*

5 *Carefully locate and cut out a portion of the soffit where the window is to be placed. Frame and insulate to suit.*

6 *Assemble the support wall to the floor extension. Make sure you overlap rough sill joints for a strong assembly.*

Next, mark the interior wall for cutting. It will be narrower than the exterior's rough opening. If the wall is drywall, cut it with a utility knife to minimize drywall dust.

If the top of the window will set into the house's soffit, you will need to insulate the area above the window before installing the window. If the window is being installed on the lower level of a two-story house, you will have to construct and insulate a custom canopy to cover the window.

For a window of this size you will need three additional people to lift the window into place. Place the plastic side flange under the siding and insert the window. While the others hold the window in place,

7 *Remove 1 in. more of exterior siding and test-fit the supplied side flanges. They should fit under the siding.*

8 *Install the window, and drive shim under the sill so the window is flush with the existing soffit. Level the window.*

9 *From the house's interior, take a small pry bar and snap the side flange into the window's groove.*

drive shims under the sill to lift the window up against the soffit. Make adjustments so that the unit is level and the windows open and close properly. Then secure the top ¾ in. plywood platform to the soffit and header. Drive nails or wood screws from under the rough sill through the shim and into the window sill.

Snap the side flange to the window with a small pry bar. Then nail the side flange in place.

Custom cut and install the nailing block and the supplied auxiliary casing.

Finishing Touches

Install outlet boxes, and run and hook up your electrical wiring. Then insulate the walls and flooring. Cut ½ in. plywood to fit under the joist extension. Install sheathing and siding to suit your house design. Then caulk and paint the exterior. Stuff insulation around the window's interior gaps, but do not over-compact.

Nail a floor patch to level and fill in the space between the existing floor and the bay window floor. Fill in all gaps with an appropriate waterproof wood filler.

Staple plastic to the inside wall to serve as a vapor barrier. Then drywall or panel (if appropriate) the wall.

Paint the drywall and window and install the matching flooring.

Cut and fit the floor and window

10 *Cut a nailing block and install the supplied jamb. It should fit up to the window's upper plywood section (top platform).*

moulding and jambs; tack in place. Then remove and finish them before affixing them permanently. Remember to place a bead of caulk on the sill before installing the sill trim. Pre-drill all nail holes to avoid splitting the trim. — *by Al Gutierrez*

Sources:

Bay window (45-3452-20): Andersen Corporation, Bayport, MN 55003.

Lumber, flooring: Knox Lumber, a subsidiary of Payless Cashways, 801 Transfer Rd., St. Paul, MN 55104.

11 *Cut the lower window trim to fit. Use a saber saw to custom cut the trim's shape.*

12 *Install exterior sheathing, siding and trim. Paint and/or stain to suit. Caulk all joints.*

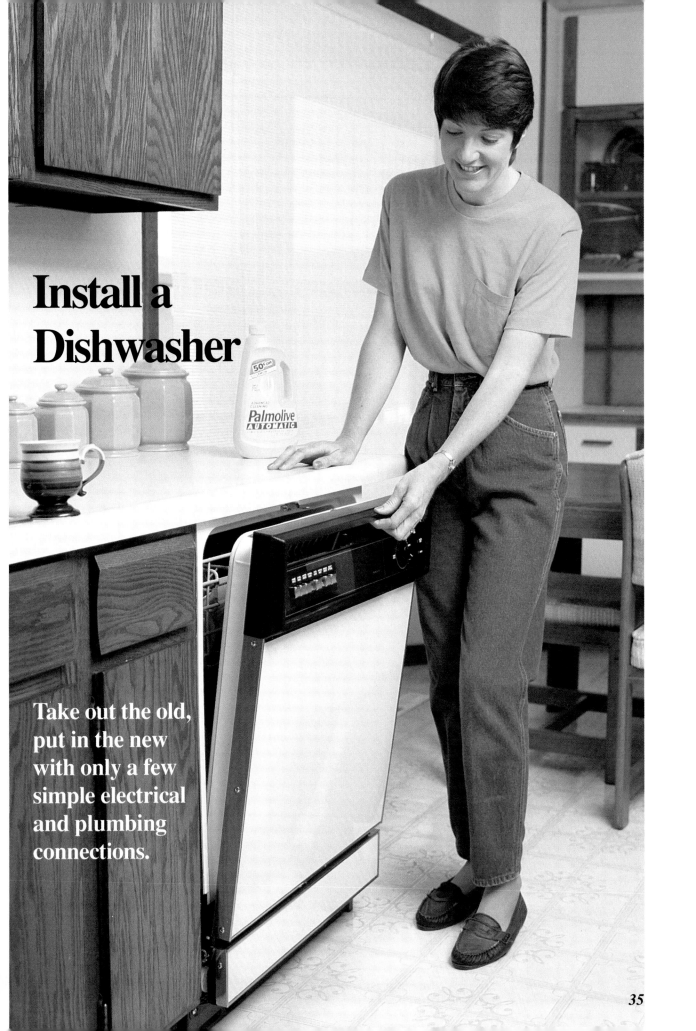

Install a Dishwasher

Take out the old,
put in the new
with only a few
simple electrical
and plumbing
connections.

Are the steel basket wires of your dishwasher broken or exposed? Do you hear the rumble of the motor even when you're outside? Does only one cycle actually wash your dishes? It may be time to replace that old and worn dishwasher.

Like so many built-in appliances, it's just a matter of time before the dishwasher needs to be replaced. To help you tackle the job of replacing a worn unit, we have tips as well as installation information.

Where to Begin

Read the dishwasher installation instructions as each unit is slightly different. Then turn off the water line to the present dishwasher. The shut-off valve is usually located under the kitchen sink. Most importantly, *locate the circuit breaker that operates the dishwasher and turn it off.* If it is difficult to locate the dishwasher circuit breaker, have a helper turn on the dishwasher, and then signal when the correct breaker is flipped off.

Note on the panel box which breaker operates the dishwasher. Leave it turned off for the entire installation process.

Remove the Old Dishwasher

With the water and electricity off, begin by removing the dishwasher drainage hose that connects to the garbage disposal or sink plumbing. Simply remove the adjustable clamp from the hose.

Now remove the two screws that secure the unit to the underside of the countertop. The screws are at the front of the dishwasher (see photo).

Carefully pull the dishwasher out from the counter cavity. If the dishwasher is located at the end of the countertop, as was the case in the photographed installation, it may be easier to also remove the side panel to better view the work.

Pull the dishwasher forward, looking for a copper pipe that connects to the dishwasher. This is the water supply pipe. Be careful not to pull the dishwasher too far out so as to bend the pipe. Use a straight

wrench to remove the supply line from the dishwasher.

Now pull the dishwasher as far out as possible, checking for hangups as it is pulled. Look for the junction box which connects the electricity to the unit. Disconnect the electrical line from the dishwasher. *Again, the electricity must be turned off before you begin.* Remove the wire nuts and remove the clamps holding the main electrical line to the junction box.

Totally remove the dishwasher, pulling out the drainage pipe. Be careful not to scratch the kitchen floor.

Install the New Dishwasher

The new dishwasher comes with a drainage hose, but it may not be long enough for every installation. Test the hose length, then decide either to use the old hose, use the new one or use the new one with a patch that lengthens it. Next, attach the drainage hose to the new dishwasher with the clamp supplied.

Locate the inlet where the water supply line connects to the new dishwasher. It may be necessary to install an elbow before connecting the water supply. The instruction manual clarifies this step.

If an elbow is needed, it may be possible to remove the elbow on the old dishwasher and install it on the new one. If so, apply plumber's putty to the elbow threads, and install the elbow on the new dishwasher.

Now feed the drainage hose back through the ports cut in the cabinet, and begin pushing the dishwasher into the counter cavity.

The type of dishwasher being installed determines

It rumbled, groaned, washed dishes on one cycle only, and lost most of its basket wires before the owners installed a new dishwasher.

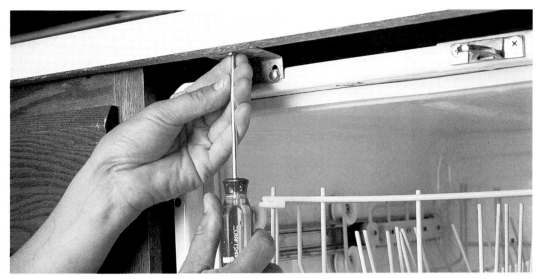

1 *Remove the two screws that secure the dishwasher to the countertop. These are the only fasteners that hold the dishwasher in place.*

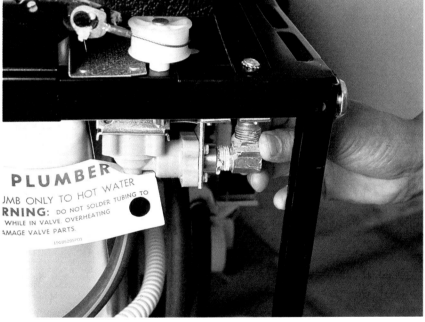

2 *Remove the 90 degree elbow from the old dishwasher and install it on the new unit. Apply plumber's putty to insure a tight seal.*

3 *Connect the dishwasher's drain hose to your garbage disposal unit.*

which line is connected next — the water supply line or the electrical line. Follow the installation manual.

Connect the electrical line to the dishwasher junction box, and use wire nuts to connect the two hot lines (black and white colored wires). Connect the ground wire to the junction box as shown in the manufacturer's installation manual.

Next, connect the water supply line to the dishwasher. Then continue to push the dishwasher into place, being careful not to kink the water supply line. Move the unit just far enough to allow inspection of the dishwasher for leaks.

Now, turn on the water supply line and check for leaks. Run the water five minutes to be sure the connection is good. Then turn on the circuit breaker and run the dishwasher. Check for leaks. If there are leaks, you will find them at the water supply line hookup, or at the location where the drainage hose connects to the sink plumbing. The solution may be to tighten the clamps or line connectors.

When the connections are tested and are leak-free, push the dishwasher all the way in; secure it to the countertop with wood screws. Run the dishwasher again with a load of dirty dishes and check again for leaks around the dishwasher. — *by Al Gutierrez.*

Sources:
Kenmore Dishwasher (22AP14795N):
Sears Catalog.

Mix custom stain colors by combining varying amounts of stains from the same manufacturer. Never mix two different brands, and be sure to write down your formula.

Enhance the Natural Beauty of Your Wood

The stain and its application are as important to your project as the wood you choose.

So you have some wood you are ready to stain — Aunt Agatha's old maple dresser, or the living room trim in your new house, for example. Understanding some basic principles about wood stains and how to work with them will go a long way to contributing to the success of your refinishing project.

When choosing a stain, always remember the very important first guideline to successful wood staining: Choose a stain that brings out the natural beauty of the wood.

Each wood variety has its distinctive character created by grain pattern, pore size and color. Cherry has a reddish hue, while pine has a yellow tint. Birch is a very light colored wood, while mahogany carries rich burgundy tones. Oak appears in a range of golden shades, and walnut has a reddish-brown look.

While wood stains come in a wide array of wood-toned colors, the happy wood refinisher — from first-timer to experienced professional — understands that wood stains will enhance the distinctive features of wood, not disguise them.

Wood stain will never help you change Aunt Agatha's maple dresser into a walnut vanity. Trying to do so will only result in a muddy and unattractive surface. A successful wood stain will, however, accentuate the graceful flow of the maple grain of Auntie's old dresser.

This story is courtesy of the Minwax Company, Inc.

The most popular types of stain used by refinishers are oil-based liquids containing dyes and colored pigments. When these stains are applied to wood, left to stand anywhere from a few seconds to 5 minutes, and then wiped off, microscopic remains of the pigments stay in the wood and affect the color.

With a little planning and some pre-testing, you can experiment to discover a stain that highlights the natural beauty of the wood in your project.

When choosing a stain color, carefully study the color charts. Many companies provide store displays that feature color charts on different types of wood. If not, bring in a small wood sample and ask the clerk for assistance in trying out stains.

If you want to experiment, you can also mix your own stain color. Be sure to use stains manufactured by one company, since it is not advisable to mix products from different lines. Buy only very small cans, and experiment with a mixture to achieve just the look you want.

Remember to keep track of your formulas in a notebook. You don't want to discover too late that you've lost track of the formula for your masterpiece stain creation. Experienced refinishers will tell you to never trust your own memory. Write down the formula as you proceed.

Sanding Techniques

1 *Use rounded objects wrapped with sandpaper for curved surfaces or a strip of sandpaper for spindles or dowels.*

2 *Wrap a small block of wood to make a sanding block for flat surfaces.*

Three Ways to Apply Stain

1 *Use a paintbrush and move the brush with the grain.*

2 *Apply stain with a cotton rag and then wipe away the excess stain with another cotton rag.*

3 *You can also use disposable foam brushes to apply stain.*

Remember, always test in an inconspicuous place.

Staining too dark is one of the most common mistakes. Trying to make a wood darker than it can naturally go can result in a muddy look. Since it is difficult if not impossible to reverse a stain, pretesting is an important step.

Another factor that affects the look of your stain is the amount of time you leave it on. Again, use a sample board or an inconspicuous spot to experiment with the amounts of time the stain is left on before being wiped off. And don't forget to keep a record so you know just which time period you want to use.

Three Ways to Apply Stain

Understanding some principles about applying stains can also help your project proceed smoothly. Three basic methods help you get the stain from the can to the board. All three methods require rubber gloves.

You can use the two-rag method. Use old absorbent cotton rags. Apply the stain with one rag, wipe away with the other. When the first rag becomes too saturated, promote the wiping rag to the application step. Old cotton athletic socks worn over rubber gloves make excellent stain applicators.

Just a word about the disposal of these rags. Before disposing, lay the rags out flat, or hang them over the side of the garbage can, to allow evaporation of volatile oils. These rags can be dangerous if wadded up and thrown in a corner.

The second method uses a normal paintbrush to apply the stain and an old rag to wipe it away. While the paintbrush keeps hands clean of stain, a drawback of this method is the unavoidable splatters, especially on upholstery.

The third method uses disposable foam brushes for applicators and rags to wipe away the stain. This is an excellent way to apply stain. Inexpensive and available in different sizes, foam brushes can be loaded with a good portion of stain without splashing and are great for applying stain to corners, crevices and carvings.

Whether you use rags, brushes or foam brushes, be careful not to rush. Apply the stain across the grain so that it catches in the pores of the wood. As you wipe off the stain, move with the grain to avoid leaving any streaks or marks.

Work on one section at a time, and work with complete boards.

Apply stain to one side of a bookcase, for example; don't work with half a board and then turn to another side.

Perhaps you are seeking the golden glow of oak, the burnished burgundy of mahogany, or the light tones of pine. By following these simple guidelines for stain selection and application, you will succeed in enhancing the unique beauty of the wood in your next refinishing project. — *by Carol Goodale and Bruce Johnson.*

A*pply a wood conditioner 15 minutes before staining to make stain application easier. You get smoother coverage and better results.*

How to Install a Garage Door Opener

Today's openers are simpler than ever to install! Put one up in about 3 hours.

Those who have lived with a garage door opener give testimony to the convenience that it adds to their lives. If you are thinking of installing a garage door opener for the first time but are not sure whether you can tackle the job, follow these steps to understand how straightforward installation really is. It no longer takes an experienced do-it-yourselfer to install one.

There are two types of residential garage door openers: chain drive and screw drive. Their motors, which are (1/4 hp, 1/3 hp or 1/2 hp), are capable of opening almost any residential garage door — sectional, one-piece trackless or tilt-up. The motor size needed depends upon the size and material of the garage door. Most manufacturers' literature and packaging define the opener's capabilities, which simplifies product selection.

This installation uses a Stanley Premier Model 3500, based on its reputation for safety, reliability and ease of installation. The Premier 3500 costs about $160; other Stanley models cost from $105 to $175.

The Premier 3500 is a chain drive model powered by a 1/2 hp motor. Its heavy-duty lifting power is enough to open a door up to 7 1/2 ft. high by 18 ft. wide. This opener includes two built-in safety features — a light time-delay that stays on 4 1/2 minutes after the door is activated, and an on/off pull cord that changes the 200 watt light into an overhead worklight. This Stanley opener also includes 1,024 changeable digital transmitter codes (so your garage opener will respond only to your transmitter and not your neighbors'), a vacation switch that shuts off the opener for an extended time, and Signal-Block, an exclusive security feature.

The battery-powered handheld transmitter that opens the garage door has two pushbuttons. One of the buttons opens and closes the door, the other activates the Signal-Block feature that prevents stray radio signals from opening the door.

Installation requires only a few hand tools and takes an average of 3 hours. An easy-to-follow, illustrated owner's manual provides clear instructions and plenty of helpful tips. A videotape included with the opener shows an actual installation.

1 *Begin assembly by connecting the tube sections. Simply insert them together, making sure the tube sections are fully seated.*

2 *At one end of the tube, insert the plug button.*

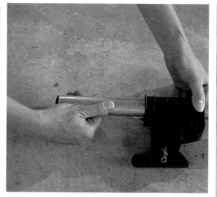

3 *Slide the traveler onto the assembly, making sure the arrow on the traveler points toward the door and away from the power unit.*

4 *Insert the end of the tube assembly with the plug button into the tube support.*

5 *Position the traveler 50 in. from the front edge of the tube support.*

6 *Attach the chain to the traveler using a master chain link.*

However, not all garage door opener manufacturers have videotape installations available.

Preparation

Before beginning assembly, view the manufacturer's installation videotape and read the owner's manual. Studying the assembly and installation steps first saves a considerable amount of time. Also have on hand the tools which are needed.

Assembly

First, join the five sections of steel tubing (Fig. 1). Make sure the tube sections are completely seated. Then press the plug button into the end of the tubing (Fig. 2). Slide the traveler on the tube assembly so the arrow on the traveler points toward the door (Fig. 3). Push the idler assembly onto the end of the tube assembly. Next insert the tubing end with the plug button into the tube support on top of the powerhead (Fig. 4).

Measure 50 in. from the front edge of the tube support (Fig. 5), and accurately mark the tubing for future reference. Position the back of the traveler at this mark. Next attach the loose end of the chain to the traveler, using a screwdriver to push the master chain link's retaining clip in place (Fig. 6). Carefully unroll the chain and wrap it around the drive sprocket on top of the power unit (Fig. 7). Thread the cable at the end of the chain through the idler assembly (Fig. 8). Making sure that the traveler hasn't moved from the 50 in.

mark, attach the end of the cable to the traveler with the master chain link (Fig. 9).

Tighten the chain by turning the adjustment bolt on top of the power unit. Tighten the bolt until the chain sags about ½ in. below the midpoint of the tubing assembly (Fig. 10). The final assembly step is to tie the manual disconnect cord to the latch on the traveler (Fig. 11). This safety feature makes it possible to disconnect the traveler and raise the door by hand in case of a power outage.

Installation

A garage door opener should be installed only on a properly operating door. If the door is difficult to open or quickly slams shut, have a professional serviceman adjust the door. An improperly adjusted door can be a safety hazard and can put an unnecessary strain on a garage door opener.

7 *Run the chain from the traveler back to the power unit and loop the chain around the drive sprocket.*

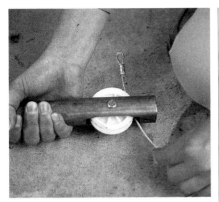

8 *Feed end of cable through the idler assembly.*

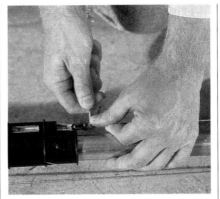

9 *Attach cable to the other end of the traveler.*

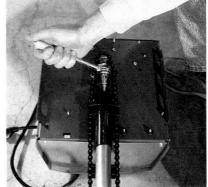

10 *Adjust the chain/cable tension so there is only ¹/₂ in. of chain sag at the tube midpoint.*

11 *Tie the disconnect cord to the latch on the traveler.*

12 *Locate and mark the center line of the door.*

13 *Find the highest point of door travel by raising the door slowly.*

Begin by measuring and marking the center line of the garage door at its top. Use a level to transfer this line to the header joist above the door (Fig.12). Raise the door to its highest point and measure the distance from the floor to the door's top edge (Fig. 13).

Next add 2 in. to the highest point measurement and mark this measurement on the header at the center line. Align the bottom edge of the bracket with this new measurement, making sure it is centered, and fasten it with two lag screws (Fig. 14). With the powerhead on the floor, lift the end of the tubing. Attach the idler assembly to the header bracket by inserting the clevis pin through the bracket and idler assembly, and securing it with a lock pin (Fig. 15).

Carefully lift the powerhead and set it on top of a ladder in preparation for hanging the unit from the ceiling joists. Position the unit so the tube assembly slopes up gently toward the header bracket and hold in place. Have a helper slowly raise the door to check the clearance.

With the proper position located, simply secure the mounting straps to an exposed joist with lag screws (Fig. 16). If the garage has a finished ceiling, attach a 1 x 6 board, using screws, through the drywall into the ceiling joists. Then screw an angle iron to the 1 x 6 and bolt the mounting straps to the angle iron.

Next bolt the L-shaped steel link to the straight steel bar link. Fasten the door bracket to the end of the L-shaped arm with a shoulder bolt. Then attach the straight steel bar link to the traveler with a clevis pin and lock clip (Fig. 17). Tug on the disconnect cord to unlock the traveler and slide it up to the closed door. Hold the door bracket against the door so that it is located even with the top two rollers (Fig. 18). On wood doors, mount the bracket to the door with two carriage bolts that are installed through the door with their heads on the outside. For steel doors, mount the brackets using self-tapping steel screws from the inside.

The final installation step is to wire the pushbutton wall switch to the power unit. Attach the wire leads to the switch and mount the switch to the garage wall (Fig. 19). For safety, the switch should be located above the reach of children and where a person pushing it has an unobstructed view of the door. Run the wire to the powerhead and connect it to the terminals on its back (Fig. 20).

14 *Fasten header bracket at least 2 in. above door's high rise dimension.*

15 *Raise the tube section and fasten to head bracket.*

16 *Hang power unit from ceiling, using mounting straps and hanging angle provided.*

17 *Attach L-link, bar link and door bracket to traveler.*

18 *Fasten bracket to door, locating it even with top two rollers.*

19 *Install pushbutton switch inside garage, being certain to mount it out of the reach of children.*

Adjustments

The final adjustments to an automatic garage door opener are critical to insure smooth, safe operation.

First, set the same radio frequency code on the transmitter and the power unit. Push the dip switches located inside the transmitter and on the back of the power unit (Fig. 21). Use a sharpened pencil or other pointed object to randomly set the switches.

Next plug the opener's power cord into an electrical outlet. If an outlet isn't within reach, have a licensed electrician install one near the unit.

Test the reverse force adjust-ment by pressing the transmitter button to close the door. While it is closing, use both hands to obstruct the door, making sure not to stand in its path until it reverses automatically. If the door is difficult to stop, turn the large sensitivity knob on the underside of the powerhead counterclockwise a quarter-turn to increase sensitivity. If the door reverses when barely touched, then twist the knob clockwise a quarter-turn to decrease its sensitivity. Repeat this procedure in quarter-turn increments until the door is properly adjusted. Always keep the door adjusted for maximum sensitivity.

Now adjust the open and close positions of the door, beginning with the close limit. Push the transmitter button to close the door. It will stop a foot or two above the floor. Slowly rotate the small black knob on the underside of the powerhead clockwise until the door is completely closed (Fig. 22).

Press the transmitter button to open the door. It will stop before it is completely open. Slowly rotate the small white button on the underside of the powerhead counterclockwise to fully open the door.

Next try the safety reverse function by placing a 1 in. thick board on the floor across the threshold (Fig. 23). Push the transmitter button to close the door. The door should

20 *Run wire from pushbutton switch to terminals on the back of the powerhead.*

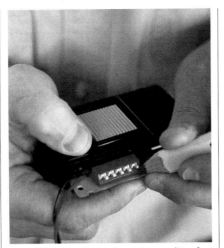

21 *Set security code on transmitter by moving dip switches; match switches with powerhead.*

22 *Set close position by turning small black knob clockwise. Door will inch down to its full-close position.*

Garage Door Opener Safety

☐ Teach children that the garage door opener system is not a toy! Do not allow children to play with it. Make sure that the wall button is mounted high enough that a small child cannot reach it. Do not leave hand transmitters accessible to children.

☐ Do not operate the door unless it is in full view.

☐ Never pass under a moving garage door.

☐ Know how to operate the door opener emergency release mechanism and test it on a regular basis (once a month is recommended).

☐ Test the door safety reverse at least once per month, according to the manufacturer's instructions. A garage door opener manufactured prior to April 1, 1982, should reverse off a 2 in. high solid object placed on the floor under the center of the door. An opener manufactured after April 1, 1982, should reverse off a 1 in. high solid object. (A section of a 2 x 4 is good for this test.)

☐ If an opener fails to pass the test, discontinue its use, and consult a professional serviceman, or contact the manufacturer. Any door opener system that cannot pass the above test after service should be replaced.

☐ A garage door opener is not a solution for a garage door that is difficult to open or close. An improperly adjusted door is a safety hazard. A professional serviceman should repair the door prior to attaching a garage door opener.

23 *Check the reverse function, making sure the door will reverse off a 1 in. block.*

reverse after coming in contact with the block. If it doesn't, slowly adjust the close limit knob until the door properly reverses. This feature of the garage door opener should be checked monthly to insure safe operation.

With the installation properly completed, you will enjoy years of convenience with your new garage door opener.

For more information, write to Stanley Home Automation, 41700 Gardenbrook, Novi, MI 48375. If questions arise concerning installation, call (800) 521-5262. — *by Michael Murray, Dean Johnson, Joanne Webeler, Dana Lowell.*

Wood Path Lights highlight your walkway and shrubbery.

Deck Lights highlight every step and rail.

Low-voltage lighting emphasizes architectural beauty.

Dockside, Post Lighting offers an extra edge of beauty and safety.

Beautify Your Landscape with Outdoor Lighting

A snap to install, low-voltage lighting adds a glow of beauty to your yard, plus extra security. Here's a roundup of the brightest ideas in landscape lighting.

Installation Details

1 *Mount the power pack transformer and plug into a standard 110-volt outlet (indoors or outdoors). For safety reasons, hang the pack at least 4 ft. above the ground. Make sure the cord can reach the outlet. Do not use an extension cord.*

2 *Place the photoelectric cell outdoors to properly turn your lights on and off at dusk and dawn.*

If you want to know the extent of America's passion for home-based outdoor living, just take a drive around your neighborhood. You'll see all the symbols — backyard decks, patios, gardens and compact greenhouses. You'll also see the newest and fastest-growing trend in landscaping and home leisure: outdoor lighting.

Outdoor lighting isn't really new. Up until a few years ago, most of it consisted of backyard floodlights or front yard lamp posts. But today's version of exterior lighting is far more attractive, versatile, safe to use and easy to install. Better yet, it's easy on the family budget. The latter ties in well with the term "low-voltage lighting," the most apt description for the new generation of "lightscaping products."

Thanks to savvy and innovative manufacturers who carefully watch consumer trends and attitudes, low-voltage lighting is now available in a wide range of sizes and styles. And these lighting products are designed for do-it-yourself installation.

According to Rob Beachy, marketing manager for The Toro

Company's Home Improvement Division, outdoor lighting falls into two basic categories — functional and aesthetic.

Functional outdoor lighting includes lighting for home security and safety purposes. A row of security lights lining a front sidewalk or driveway will discourage most intruders. Also, it is a fact that our population has a higher proportion of older people. Well-placed safety lights illuminate any steps, sidewalk borders or other obstacles that present potential hazards. Although swimming pool lighting is more aesthetic in intent, it is certainly a safety factor for people of all ages.

"Aesthetic lighting is primarily used for looks, focusing on a feature of your home to emphasize its architectural beauty. Many homeowners use smaller 'mood lights' to accent a garden, tree or statue," notes Beachy. "And owners certainly use low-voltage lighting for patios, pools and decks during the evening, for aesthetic as well as functional purposes."

Experts at the Minneapolis-

based Toro Company point to another big factor in the popularity of low-voltage lighting: home value. Real estate professionals say attractive landscaping can add as much as 15 percent to a home's value, and homeowners can get as much as a 200 percent return on their investment in landscape plantings. That fact well illuminates the value of low-voltage lighting. "Lightscaping" these outdoor assets can increase the landscaping payback even more. When selling your house, prospective buyers will see low-voltage lighting as a real bonus.

Ease and safety of installation may be the strongest reason for the spiraling interest in low-voltage lighting. The term "low-voltage" means just that. Today's lighting is designed with 12-volt safety for do-it-yourself installation. The lights operate off of your normal household 110-volt current. Toro, for instance, features a power pack in its lighting kit that is equipped with a transformer. This reduces the 110-volt current to a safe 12 volts.

However, low-voltage doesn't mean low intensity. Toro's six-light

3 *With every low-voltage fixture, electrical contact is made simply by pushing the cable onto two prongs.*

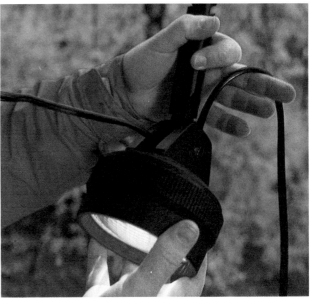

4 *String the cable through or around the fixture's stake and hide it from view.*

set of floods, one of many varieties available, provides more light than a pair of auto headlights.

Now that you are familiar with the basic facts on low-voltage lighting, let's take a close-up look at some of the factors that most people wonder about.

Cost

Basic lighting packages of four to six fixtures, including all necessary installation materials, start at about $50. You can begin small, then add on later. Operating costs are minimal. A set of six lights, running on a 12-hour, dusk-to-dawn schedule, will add only pennies a day to your monthly electric bill.

Easy to Install

Most low-voltage lighting kits don't require tools. If you can tell a cord from a fixture and identify an electrical outlet, you can install a low-voltage system.

In simplest terms, the lights snap together for easy assembly. Electrical cords needn't be buried for safety, though you can bury them for appearance sake, or for lawn-mowing convenience. Lights fit onto

stakes that are pushed into the ground. Extra lights can be added by attaching additional cable, with a small cable connector, anywhere along the existing cord.

Durability-Plus

Most low-voltage lighting systems are made of high-impact, weather-resistant, engineered plastics. For instance, Toro lights are designed to withstand temperatures from 35 degrees below zero to 120 degrees above zero. Fixtures won't rust, even near the ocean. Also, the power boxes are totally weather-proof.

Low-voltage lights can be operated manually or automatically. On the automatic side, you can have a dual-position timer that goes on at dusk and off at dawn. Or, you can set a variable timer. Lights also can be controlled by an outdoor darkness sensor, or photo cell, that comes standard with any timer. The sensor has a delayed-reaction feature so stray lights flashing on the sensor won't accidentally disrupt the system. — *by Ray Lorenz. Photography by The Toro Company.*

5 *Finally, stick the light into the ground, mount it on a tree or attach the deck mount to a wood surface. Bury the cable under turf or cover it with mulch or dirt, as shown here.*

Redo Your Kitchen Floor

Give your kitchen a new look with easy-to-install floor tiles.

After years of stomping feet, crashes and spills, the Clarks' rust-colored flooring was worn and difficult to keep clean. They installed new Armstrong flooring over their 10 ft. by 20 ft. kitchen/dining room floor in two days.

What to do with the Old Flooring

Laying tile flooring goes quickly, and you can immediately see the fruits of your labor. The most tedious work comes with the removal of the old flooring, and the scraping of the adhesive residue. However, you do have an alternative.

If there are no other floor layers under your old floor tiles and if the existing floor covering is not severely embossed, you can keep the old flooring in place (below photo). In this case, patch any holes, gaps, etc., with a latex patching compound. You will still have to remove moulding. (Where new floor covering meets any adjoining flooring, plan to install an appropriate threshold if possible.)

If there is another layer of flooring underneath, or if the flooring is deeply embossed or uneven, etc., you will have to remove the flooring.

Where to Begin

Whether you remove your old flooring or opt to keep it in place, first remove the wooden base moulding with a small pry bar. On long pieces, insert the pry bar every 2 ft. and work the moulding away from the wall. Work evenly. Trying to pull the moulding away in one fell swoop can result in wall indentations (from the pry bar) and broken moulding. Mark the wall location of the moulding on the back of each piece to ease reinstallation.

Next, with a sharp utility knife, cut along perimeters where no base moulding was installed, such as along kitchen cabinets. Lay on the floor and use two hands to control the cut. Tilt the utility knife at about 5 degrees from vertical to avoid cutting into the base of the cabinets.

Also, cut along the floor threshold that divides the tile from carpeting or other adjoining flooring. Remove the trim if you are going to repair the carpet or reuse the threshold. Remove any other thresholds that divide the tile from other flooring, such as oak flooring.

Tile Removal and Preparation

Starting from one end of the kitchen, heat up the corner of one of the tiles with a heat gun and insert a 3 in. stiff putty knife under the tile. Heat up the area ahead of the putty knife and push the blade. The tile should begin to lift and come off. After a few tiles you will have removal down to a science.

If you work from all the walls inward, you will find yourself kneeling on old, sticky floor adhesive. So,

This old floor was in pretty poor condition, but was left in place.

1 *Remove base moulding with a small pry bar. Place a piece of masking tape under the bar's foot to avoid wall scratches. Then pull the finishing nails from the moulding's back with locking pliers. This reduces face blemishes.*

2 *If you decide to remove the old flooring, cut the existing tile at thresholds, kitchen cabinets, built-in appliances, etc., with a sharp utility knife. If the threshold can be replaced, do so.*

3 *If the old floor tile must be removed, peel off each tile with a heat gun and a 3 in. putty knife. Heat a tile end, insert the putty knife and push the knife forward. Wear gloves.*

4 *Fill in all holes and dents in the underlayment with a good waterproof wood filler. If you leave the floor in place, use a suitable filler available at flooring retail stores.*

always work towards another open room area.

Remove the existing floor adhesive with a floor scraper. This tool can be rented from tool rental stores or perhaps through the store where you purchased the tile flooring. This method is labor intensive and slow, but it will remove about 80 percent of the floor adhesive residue.

Patch all holes and indentations with a plastic wood filler. Then sand these patches.

Finally, vacuum up all debris from the floor. Even a speck of old tile will form a bump in your new flooring.

Layout and Floor Application

You must lay out a starting grid to insure the new flooring will run parallel to the walls and minimize tile waste. You do not want to have a row of tile that is 1/4 in. wide. If you are applying the same size tiles on a floor that you removed, you can use the old grid. (The old chalk line may be visible and the tile outlines appear as stains on the underlayment.) However, if you are tiling over an existing floor, plan the layout so that the seams of the new floor will overlap and not coincide with the old seams.

Sources:

Century Solarian (24612) flooring: *Armstrong World Industries, Inc., P.O. Box 3001, Lancaster, PA 17604.*
Table and Chairs (1B2545N): *Sears Catalog.*
Wood filler: *Minwax Company, Inc., Dept PS-WB, 102 Chestnut Ridge Plaza, Montvale, NJ 07645.*

Think of the kitchen as a rectangle, with grid lines dividing the longest and shortest widths in half. Snap new chalk lines after having used a tape measure to determine the starting grid lines.

If you are applying a tile of a different size, determine where (from the starting grid) to begin the flooring. Use the old line as a reference and snap new parallel lines.

From here on, you should work on laying a quarter of the kitchen (or less) at any one time. Remove the release paper from the back of a tile and press the tile into place. Use hand pressure to secure the tile. Work by making a progressively

5 *Determine the grid line locations and snap two chalk lines. If the old floor has been removed, use the old chalk line as reference.*

6 *Peel the release paper from the back of the tile and lay the tile in place. Rub it firmly with your hands for a good bond.*

larger square area. Lay in the next tile by pulling it in tightly to the adjoining tile at one end, then pressing it down. Tile installation goes quickly.

Lay the remaining tiles (for this quarter), and then measure and cut tiles to suit the perimeter. Cut these tiles with heavy shears or a utility knife.

Finally, roll the floor with a floor roller or rolling pin to insure proper floor bonding. Slowly work the roller lengthwise and then from the side. Install the remaining floor in quarters, rolling each section as it is finished. Then reinstall the moulding to complete your new flooring's installation. — *by James Guthrie.*

7 *Use a floor roller or a rolling pin to create a good bond. Work in small overlapping areas.*

Home Sweet Home Improvements

Home improvement projects can have both immediate and long-lasting benefits when you consider your family's comfort and the increased value of your home.

Home improvements ranging from gleaming contemporary to country rustic decor have become one of the symbols of the decade. "Home is where the heart is" continues to be the nation's prevailing theme.

Our fever for home remodeling, repair and fix-ups has created a $100 billion a year industry that seems to be digging in for the long haul through the 1990s and beyond.

Two of the largest trade groups in the housing industry — the National Association of Home Builders (NAHB) and National Association of the Remodeling Industry (NARI) — both report that recent years have ushered in amenities unheard of 20 years ago. "Homeowners are installing custom windows and skylights, spacious bathrooms attached to bedrooms, and luxury baths with saunas and expensive fixtures," says Tom Irmitar, president of a St. Paul, Minnesota-based remodeling firm and a local director of NARI. "Upscale luxury is commonplace today, and prices for a remodeled kitchen or bath can easily exceed $50,000," he adds. "Basically, today's homeowners want total comfort, more security, space for leisure, and they're willing to pay the price. Homeowners want to tailor a home to their needs, plus make the investment pay off in higher resale value."

Define Your Objectives

If you are thinking of home improvements, a good approach is to evaluate your remodeling strategies from a basis of objectives. Is the remodeling to make your home more comfortable for you or mainly to increase your home's resale value?

The answer to this question will help you evaluate large and luxurious additions — such as an in-ground swimming pool or a tennis court — which may not pay back the entire investment, but may be worth the price to you.

Some of the improvements that you may put into your home for your own comfort pay back better than others. *Remodeling* magazine reported in its "Cost vs. Value" report that kitchen remodeling ranked highest in resale value, with homeowners regaining 84 percent of their remodeling costs when selling a home. Bath rehabs/additions and fireplace additions also scored high.

"Resale value soars when adding a bathroom to a one- or two-bath house," notes Chuck Van Eeckhout, president of Van Eeckhout Building Corp., Long Lake, Minnesota. "Today's homeowners prefer amenities such as whirlpool baths, enlarged and extra-luxury kitchens, and hardwood floors. Another strong trend is an expanded master bedroom, preferably on the main floor."

Doing It Yourself vs. Hiring a Professional

When you have decided to undertake a remodeling project you will need to decide if you will handle your project do-it-yourself style, completely or partially, or hire a professional remodeling contractor to do it all.

Several important factors should be weighed before launching any do-it-yourself home repair or remodeling:

- Do you have the hands-on ability to complete the project and have it pass local building codes?

- Do you have the time to spare? Do you enjoy this kind of project? As many have experienced, when a how-to book or instruction manual says the job will take an afternoon, in actuality it might take considerably longer.

- Will you be able to save money on your do-it-yourself project? Do you have the tools; can you purchase materials reasonably?

If your project will be a major addition or a complete room renovation that entails heavy demolition or potentially dangerous work on electrical or heating systems, for example, you may be wise to hire an experienced professional.

A professional generally has a better eye for design and detail. He or she is educated in reading blueprints, can usually get the job done more quickly and within a specified time. A licensed pro has to make sure the final project isn't out of plumb, doesn't leak and is up to code.

Codes and Permits

Building codes are standards set by your municipality and you must meet these standards and pay the appropriate inspection fees.

A building permit generally is required whenever structural work is involved or when the basic living area of a home is to be changed. Your local government offices can provide you with all the information or a contractor can get the required permits for you.

Financing

An important part of your planning will be financing.

When you're planning a major remodeling project with a price tag of $10,000 or more, a loan is usually required. Here are a few alternatives to consider:

☐ Personal Loan: Usually called a "Home Improvement Loan," it's the most widely used type of loan. If you're a regular customer with a top credit rating, collateral sometimes isn't required. Interest rates tend to be higher than a conventional mortgage. The length of the loan? About two to eight years.

☐ Contractor's Loan: Here, the contractor joins hands with a bank. He will arrange a loan for the estimated cost of the project, and the interest rate is usually higher than other loans. The loan is usually for about one to five years.

☐ Home Equity Loan: A home equity loan can be structured in one or two ways. First, it can be structured as a traditional second mortgage, wherein the borrower obtains funds equal to the full loan amount immediately and commits to a fixed repayment schedule. Alternatively, home equity borrowing can be structured as a line of credit, with check, credit card or other easy access to the credit line over its lifetime.

HOW VALUABLE IS YOUR HOME IMPROVEMENT?

When you remodel your home, it usually isn't solely for your own comfort. If you're like most homeowners, you hope the time and money spent on a project will also increase the value of your home. Here's an abbreviated sampling of the "1988 Cost vs. Value Report," an annual tally of costs and pay-back values for 14 typical remodeling projects. The survey is conducted every year by *Remodeling* magazine, a publication for contractors and architects.

Major Kitchen Remodeling

Includes design and installation of a functional layout of new cabinets, countertops, energy-efficient appliances and custom lighting, new floor, wall coverings and ceiling treatments.

Cities	Remodeling Cost	Resale Value	Pay-Back
Atlanta, GA	$19,966	$17,969	90%
Chicago, IL	$21,660	$19,277	80%
Dallas-Ft. Worth, TX	$19,416	$16,000	82%
Portland, OR	$21,782	$14,800	68%

Minor Kitchen Remodeling

Job includes refinishing cabinets (with minor modifications), installing one energy-efficient cooking appliance, new countertops and cabinet hardware, and cosmetic decorating (wall covering, flooring, painting and some new lighting).

Cities	Remodeling Cost	Resale Value	Pay-Back
Atlanta	$6,302	$5,672	90%
Chicago	$7,976	$7,338	92%
Dallas-Ft. Worth	$6,716	$3,850	57%
Portland	$7,868	$5,900	75%

Bathroom Remodeling

Remodeling adds new tub, sink and commode to a 5 ft. by 7 ft. room. Also included are new vanity, tri-view medicine cabinet, ceramic tile walls (to ceiling in tub area, 4 ft. high elsewhere) and ceramic tile floor.

Cities	Remodeling Cost	Resale Value	Pay-Back
Atlanta	$5,700	$4,275	75%
Chicago	$6,743	$5,057	75%
Dallas-Ft. Worth	$5,824	$3,300	57%
Portland	$6,927	$4,000	58%

☐ Mortgage Refinancing: A totally new mortgage is another option. The main point to check out here is the interest rate you are trading from your old mortgage to the new one. A

Add a Full Bath

Add a 5 ft. by 7 ft. bathroom within the home's existing floor plan, including sink, tub, shower, commode, vanity, tri-view medicine cabinet, ceramic tile floor and ceramic wall tile (to ceiling in tub area and 4 ft. high elsewhere).

Cities	Remodeling Cost	Resale Value	Pay-Back
Atlanta	$8,165	$ 7,757	95%
Chicago	$9,658	$11,590	120%
Dallas-Ft. Worth	$8,341	$ 5,500	66%
Portland	$9,921	$ 5,000	50%

Add a Fireplace

The added fireplace is a factory-built 36 in., zero-clearance, energy-efficient model with glass doors and outside combustion air through ductwork. It has a 6 in. wide, floor-to-ceiling stone or brick facing, a raised hearth and 6 in. mantle.

Cities	Remodeling Cost	Resale Value	Pay-Back
Atlanta	$3,095	$2,940	95%
Chicago	$3,625	$4,894	135%
Dallas-Ft. Worth	$3,310	$1,700	51%
Portland	$3,853	$3,853	100%

Complete survey results are available for $4.50. Write to: *Remodeling* magazine, Allison Hoover, 655 15th St., N.W., Washington, D.C. 20005.

one or two point increase could represent a payback of thousands of additional dollars in interest over the life of the loan.

Different financial institutions may offer different rates and terms, so it is a good idea to check several sources.

Insurance

Another important factor to consider in your home remodeling is your homeowners insurance. If you build a new bathroom or kitchen, not only will it enhance your lifestyle but your home value will likely increase as well. Is your present insurance adequate? GEICO, an insurance company, specializes in counseling homeowners, and even provides a 24-hour toll-free service to help you make sure that you are insured for the full value of your home and to assist you in designing a policy specific to your needs.

To find out more about GEICO's homeowners insurance coverage, call toll-free 1-800-841-3000. — *by Ray Lorenz.*

This story is reprinted from GEICO Direct magazine with permission from GEICO Insurance Company.

SELECTING A REPUTABLE REMODELER

Finding a qualified, professional remodeling contractor is easy, but it will take some time and effort on your part. Here are some basic guidelines established by experts with the National Association of the Remodeling Industry (NARI). These how-to tips will not only make the selection easier, but you will be more prepared to make an informed decision that best suits your needs.

☐ Ask the contractor if the company is insured against claims covering worker's compensation, property damage and personal liability in case of accidents. Verify the contractor's insurance coverage with the carrier and agency.

☐ Check with state, county or city housing authority to insure the contractor meets all licensing and/or bonding requirements in your area.

☐ Make sure he or she is an established business in your local area. Local firms can be checked through past customers. As members of your community, local contractors are more anxious to build a solid reputation for service and quality.

☐ Check to see if the contractor is a member of the NARI or any other national remodeling/builder association.

☐ Contact your local Better Business Bureau to see if the remodeling contractor has an adverse file or record.

☐ Ask for local homeowner references and follow up on them. Ask if they were satisfied with all aspects of the remodeler's performance. Ask to see the finished projects.

☐ Make sure you get two or three bids for the work you need. Don't blindly accept the lowest. Ask the contractor why the bid is so low, or so high. Sometimes a higher price may be worth the cost of higher quality materials. Make sure you have a thorough understanding of all materials to be used.

For more information, write to: NARI National Headquarters, 1901 North Moore Street, Suite 808, Arlington, VA 22209. ☐

How to Repair Wood Decay

Soft, spongy rotted spots can develop where wood is exposed to moisture that can't escape. Now, wood repair systems make it easy for you to treat soft spots in wood.

Even the novice do-it-yourselfer can overcome beginner's hesitation and successfully accomplish these necessary wood repairs with easy-to-use, inexpensive wood repair systems.

Overcoming any reluctance to get started is important, since rotting wood presents an excellent example of the old adage, "He who hesitates is lost." If left untreated, the rot grows worse, destroying larger areas of wood, providing entry for insect infestation and generally setting the stage for extensive destruction. Using a few simple tools and a quality wood repair system, every homeowner can race to the rescue in just a few relay laps around the house.

The Warm-up Lap: Inspection

Before you begin the actual repair process, you'll want to take a warm-up lap around the house to identify trouble areas.

Look for damage in places where wood comes in contact with moisture, or where there are extreme hot and cold temperatures. Inspect areas where wood comes into contact with earth, leaves, masonry or moisture. Be aware of areas of peeling paint. Once the barrier of paint or finish has been broached, water continues to seep in, causing rot. Common trouble spots occur where wood meets masonry, such as at the bottom of decorative columns, or on

Check the exterior wood for soft, spongy areas that indicate rotting wood and scrape away loose, flaking paint. Dig out softened wood with a screwdriver or wood chisel and remove dirt and debris. Let the area dry out.

window sills or steps where water can't easily run off.

Four Laps to Wood Repair

While you can conduct your warm-up inspection on a cloudy, even drizzly day, you'll need dry weather to begin the repair process.

In addition to the waterproof wood filler, only a few basic tools are needed: a putty knife or paint scraper (or old household knife), inexpensive paintbrushes, old toothbrushes and sandpaper.

Lap 1: Remove the Rotten Wood

Rotting wood is like tooth decay — a soft, spongy mass fills a cavity, trapping moisture and further decay beneath. The soft, spongy material needs to be removed, all debris cleaned out and the area dried before a filling is provided.

For this first lap, the only tools needed are an old screwdriver, or even just an old household knife, and a paintbrush.

Begin by scraping away all of the loose, flaking paint. Then dig out as much of the damaged wood as possible. The goal is to remove as much dust, dirt and debris as possible. For particularly hard-to-reach cracks, common kitchen items such as a bottle opener or grapefruit knife can help. The dry paintbrush sweeps away loose debris.

Once the spongy decayed material is removed, air can reach areas where moisture previously has been trapped. As you complete your first lap around the house, removing all debris from rotted areas, the areas you first cleaned will have a chance to dry out.

Lap 2: Apply Wood Hardener

Wood hardener is a clear, quick-drying liquid that penetrates deeply into the wood, binding and reinforcing decayed wood fibers. When dry, the wood hardener acts as a barrier coat, sealing wood against further moisture and providing a strong base for the next step, the wood filler.

This story is courtesy of the Minwax Company, Inc.

1 *Apply a liquid wood hardener as a barrier coat and let dry two to four hours.*

2 *Then apply the thick, putty-like wood filler or putty with a putty knife. Allow the wood filler to dry several hours. A variety of exterior-grade fillers are available in hardware stores.*

Check again to make sure that the cavity is free from dirt, grease, oil and loose paint, and that it is completely dry. Use a brush to apply a generous amount of hardener as you move around the house, completely saturating each repair area.

Wood hardeners dry quickly. As you finish your second lap and return to your starting point, you can begin another lap, recoating each area with another application of wood hardener. After the final application, the wood hardener needs to dry for two to four hours. Allow time for each area to dry thoroughly.

Lap 3: Apply the Wood Filler

For this lap, you'll need a putty knife and a few nails and screws.

After the hardener has dried, insert nails or screws in the larger cavities to provide an anchor for the filler. Make sure these are inserted at an angle, with the heads below the level of the finished patch, so they will be completely covered by the filler. (Be sure not to use nails or screws for a patch that you intend to drill or carve later on.)

Using the putty knife, fill the crevice with the filler. As you apply this thick, putty-like material, pack it tightly to prevent air bubbles. As you finish filling the crevice, mound the filler up a little to provide a surface that can be sanded level.

The wood repair process is now complete. However, don't neglect to prime and paint your repairs. You want your work to last as long as possible.

The Last Lap: Sand and Paint

Allow from a half-hour to several hours for the filler to dry. It's important to allow adequate drying time before sanding repair spots. If you start to sand and don't stir up any dust, you've started too early.

Use the "fingernail test" — if it's easy to make an indent, your filler isn't dry yet. Use the extra time to inspect your outdoor furniture. You can use this same repair system to help preserve your outdoor furniture. Instead of cutting your picnic benches a little shorter every year, try using this wood repair system to repair rotten spots at the ends of the

3 *Sand the filled surfaces with a sanding block or pad sander.*

bench legs.

A piece of sandpaper wrapped around a block of wood makes an effective sanding block. You can purchase a sanding block at paint and hardware stores, or use a block of either 1 or 2 in. thick wood, about 5½ in. long and 3½ in. wide. A standard sheet of sandpaper cut into fourths can easily wrap around this block. File or sand the top edge and corners of the block so that it fits comfortably into the palm of your hand. After sanding the filler level and smooth, the repair is ready for primer and paint.

With this simple system of wood repair, you can now defeat dangerous soft spots of wood rot, and protect your home for years to come. — *by Carol Goodale and Bruce Johnson.*

4 *Then prime and paint the newly filled surface to complete your wood repair.*

Kitchen Plumbing Made Easy

Learn how to install and plumb a new kitchen sink.

Doing the finish plumbing for your kitchen consists of installing the sink. The double-bowl kitchen sink is fairly standard.

Either both bowls are centered on the waste pipe (Fig. 1, left) or one of the bowls is centered on the waste pipe (Fig. 1, right). If a new waste dis-

poser will be installed, it is usually located in the right-hand bowl. In this case, the wall waste pipe should be centered either on the left-hand

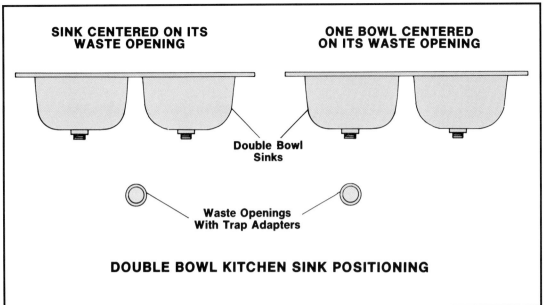

1 *Shown are a double-bowl sink with the waste opening centered between the two bowls (left) and a sink with the waste opening centered on one of the bowls (right).*

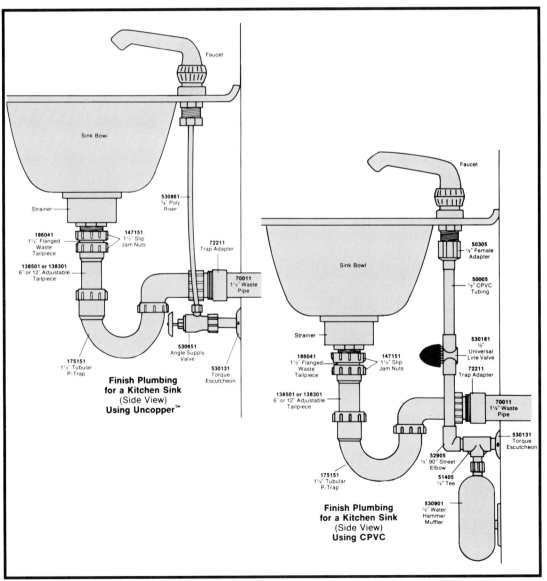

2 *Shown is a side view of all the parts involved in a sink's finish plumbing.*

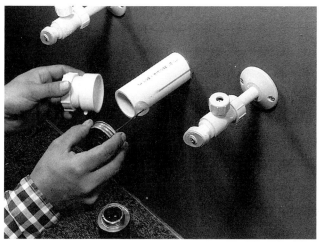

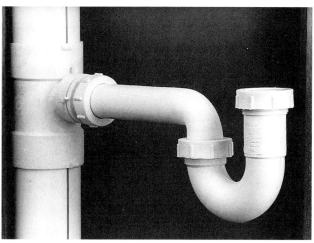

3 *Begin by cutting the capped waste pipe stubout to a 4¹/₂ in. length, then solvent weld a trap adapter (Genova Part No. 72211) to it. The water supply tubes have been cut off and fitted with Genogrip Angle Supply Valves.*

4 *If, instead of a waste stubout, you have a 1¹/₂ in. fitting socket waste opening, solvent weld a fitting trap adapter (Part No. 72311) so it will accept a tubular P-trap.*

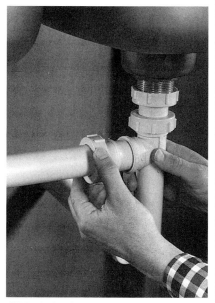

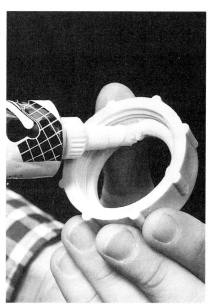

5 *Tubular drains use slip joints between parts. These consist of a slip washer and a slip jam nut to hold the washer tightly between the tube and its fitting. The fitting is threaded for the nut. A slip washer's flat side should always be installed so it faces the nut.*

6 *Pouring boiling hot water down a sink drain won't harm tubular drains if they're made of polypropylene pipes. This heat- and chemical-resistant thermoplastic is the material of choice for all tubular drains.*

7 *Here's a trick that pro plumbers often use to avoid call-backs for leaks: run a little silicone rubber sealant or plumber's putty around the inside of the face of the slip jam nut before making the joint.*

bowl or between the two bowls.

Fig. 2 shows a side view of the parts involved in a sink's finish plumbing. The first step in a sink installation is to solvent weld the trap adapter (Figs. 3 and 4).

Mount the Sink and Faucet

Most kitchen sinks are part of a counter base rather than freestanding.

They are fastened in a cutout made in the countertop. Before placing your new sink in its cutout, mount the faucet. Some faucets are set in plumber's putty, while others have their own base gaskets. Install the spray hose (if there is one) in the appropriate sink opening, following the instructions of the faucet and sink manufacturers.

The basket strainers in the drain openings are installed separately. These are sealed into the sink bowls with plumber's putty. A thick rubber gasket and flat washer go on the un-

This story is courtesy of Genova Products, Inc.

8 *Because trap bends hold water constantly, they have carefully fitted swivel joints connecting the J-bend and waste arm. Although no gaskets or washers are used here, plumber's putty or silicone rubber may be applied if the trap is metal, polypropylene or PVC plastic. A slip jam nut provides a leak-free joint.*

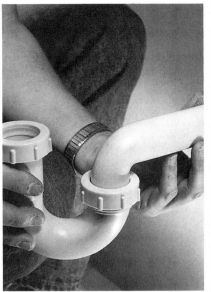

9 *A P-trap connects the sink to the waste stubout in the wall. Water is trapped in the J-bend section to seal out gases and vermin, yet permits wastes to pass through.*

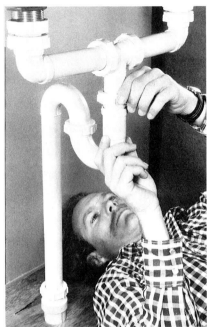

10 *Waste outlets in the floor are rare, but if your kitchen sink is plumbed this way, a Genova tubular S-trap will fit it. An extra deep trap seal prevents siphoning of trap water by the out-rushing flow. While most codes do not permit the use of S-traps in new plumbing systems, the one by Genova works fine in replacement applications.*

derside of the sink. The assembly is then squeezed tightly to the sink bowl opening with a large jam nut, which threads to the basket drain.

You can generally get it reasonably tight by hand, but to make it really tight, use a screwdriver or a special strainer wrench (also known as a

11 *Tubular drain fittings.*

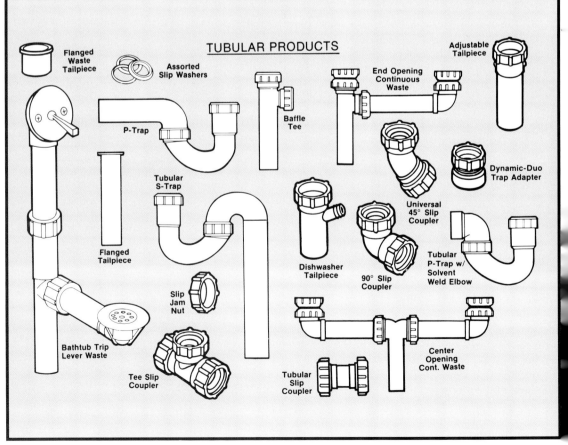

TUBULAR PRODUCTS

Flanged Waste Tailpiece

Assorted Slip Washers

Baffle Tee

End Opening Continuous Waste

Adjustable Tailpiece

P-Trap

Tubular S-Trap

Flanged Tailpiece

Dynamic-Duo Trap Adapter

Universal 45° Slip Coupler

Dishwasher Tailpiece

90° Slip Coupler

Tubular P-Trap w/ Solvent Weld Elbow

Slip Jam Nut

Bathtub Trip Lever Waste

Tee Slip Coupler

Tubular Slip Coupler

Center Opening Cont. Waste

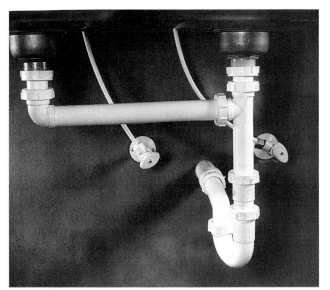

12 *An end-opening continuous waste can be arranged to fit either a right- or left-handed waste connection to a double-bowl sink.*

13 *A center-opening continuous waste is used when the waste opening is centered between the two sink bowls.*

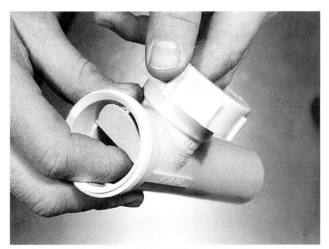

14 *A deflector plate inside a baffle tee routes disposer wastes toward the trap rather than having them back up into the unused sink bowl. This avoids having to build separate wastes for the sink and disposer. The plastic baffle may be taken out if there is no disposer.*

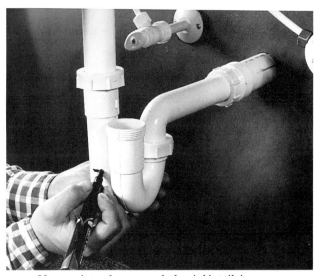

15 *If a trap is too low to reach the sink's tailpiece, an adjustable extension tailpiece will let it do so. Mark the extension for cutoff.*

spud wrench). Wipe away the excess plumber's putty inside the sink before moving on.

With the sink in place, you can now connect the water supply tubing to the faucet. It is easier to connect the water supply before installing the drain parts. If the water supply pipe is in the wall, use Genova Part No. 530651 Genogrip Angle Supply Valves and Part No. 530861 20 in. Poly Risers. These come two per pack — just the number you need for finish-plumbing a sink. If the water supply pipe comes from the floor rather than the wall, you'll need 36 in. Poly Risers (Part No. 530881).

Tubular Goods

The parts used to hook up a sink's drains to the waste pipe are called tubular products. This distinguishes them from the very different drain-waste-vent (DWV) piping. Tubular drainage products are thinner walled than DWV pipes. A tubular drain has walls 0.062 in. thick vs. 0.146 in. for DWV pipe. In other words, DWV pipe walls are more than twice as thick as tubular walls. What's more, tubular parts use slip-nut joints (Fig. 5) instead of solvent welded joints. Because tubular goods are thinner-walled than DWV, they must be made of a more heat-resistant material.

Tubular drainage products are not meant for use inside house walls. You must use them in exposed areas such as underneath fixtures where you have ready access for maintenance and repairs.

16 *Basket strainers come with flat washers that should be installed between the bottom of the basket strainers and the flanged waste tailpieces. These are soft and serve as effective leak preventers.*

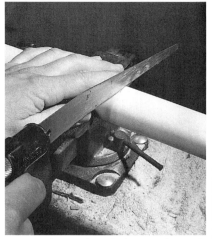

17 *Polypropylene drainage tubes and extensions may be cut with any fine-tooth saw. It helps to hold them in a vise while cutting. Make sure the pipe end is free of any burrs.*

18 *If you have a double-bowl sink with an end-opening continuous waste, fit up the baffle tee first. The baffle tee fastens to a short flanged waste tailpiece, which gets a flat washer between it and the basket strainer.*

Why Polypropylene?

Using thin-walled, tubular drainage products doesn't seem to make sense, since fixture drainpipes receive the hottest water and the most concentrated exposures to chemicals such as drain cleaners. To make the life of a tubular drainpipe even harder, hot water and those caustic, heat-producing chemicals don't flow through a fixture drainpipe as quickly as they do in a house's DWV system. Instead, hot water and chemicals stand and soak in the low points of the drain traps... for minutes, perhaps hours.

Some products, such as Genova's tubular goods, are made of a tough heat- and chemical-resistant plastic called polypropylene (PP). In fact, PP is so resistant to heat and chemical damage that it cannot be solvent welded.

Some tubular systems use PVC; it's a fine material, but not the best one for tubular drainage products.

Some systems have cheaper-to-make ABS slip jam nuts. These are threaded nuts that hold slip-jointed tubular drainage parts together and keep them from leaking. Buy polypropylene slip jam nuts. Even though you may accidentally drop some solvent glue on one, it won't be welded to its threads.

The point in bringing up the issue of solvent glue is that in installing a system, you can make use of sealants to prevent leaks in slip-joint parts (Fig. 7). But don't try this with non-Genova ABS slip jam nuts. They will fail from what plastics chemists call **stress cracking**. Stress cracking occurs when ABS is exposed to the oils in plumber's putty or the chemicals in silicone rubber sealant, two of the most common slip joint sealants. So, if you leak-proof a Genova or a metal slip jam nut, there's no problem. But with any

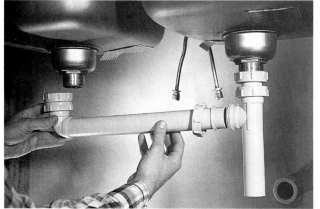

19 *The tube between the sink bowls of an end-opening continuous waste may need to be cut if the bowls are closer together than standard sink bowls.*

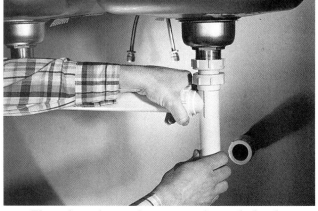

20 *The end-opening continuous waste is attached to the basket strainer at one end and to the baffle tee at the other. Tightening the slip jam nuts does the coupling.*

other type of nut, don't use plumber's putty or sealants.

Tubular Traps

The best-known tubular drainage item is the trap. Most common is the P-trap. It has two parts: a J-bend and a longer waste arm. These are connected in the center with a swivel joint (Fig. 8). The parts of a center joint are carefully designed to allow a trap to be swung to the right or left so it can be properly aligned with the fixture waste opening. P-traps are used with waste openings in walls.

Another kind of trap, called an S-trap because of its "S" shape, is used with waste openings in floors. It's the only practical way to serve a fixture with a floor waste connection. S-traps are banned from use in new construction by most of today's plumbing codes because they siphon easier than P-traps, sometimes losing their protective water seals. However, in most replacement installations, Genova's S-trap (Part No. 177151) works fine. (See Fig. 10.)

Tubular parts (Fig. 11) for kitchen sinks and laundry tubs are 1½ in. in diameter, which matches the diameter of DWV waste openings in most homes.

The least-known and hardest-to-find tubular parts are the small slip coupling fittings, including 90 degree and 45 degree elbows, tees and straight couplings. You are likely to need these only if you are designing your own under-the-counter tubular system for a double-bowl sink.

Continuous Waste

You can buy a packaged drain system that includes all the tubular fittings needed. These are available primarily for double-bowl kitchen sinks. You need one type for a sink with both bowls centered on the waste opening, and a different kind for a sink with one of the bowls centered on the waste opening. Either way, the tubular drain setup is called a **continuous waste**. An end-opening continuous waste (Genova Part No. 186001 — Fig. 12) is used when one bowl is centered on the waste opening. A center-opening continuous waste (Genova Part No.

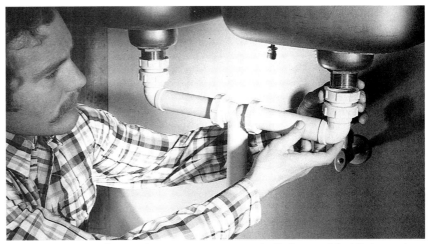

21 *A center-opening continuous waste connects similarly except that its tee centers between the two sink bowls. As with an end-opening continuous waste, use a flat washer between the flanged waste tailpieces and basket strainers.*

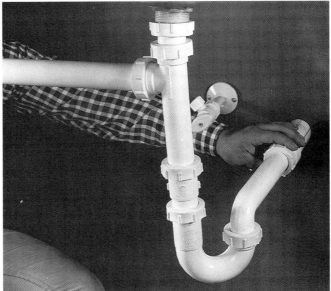

22 *A trap swivels to reach a waste opening that is offset somewhat from the sink's drain. With drain tailpiece and waste opening aligned, the trap aims straight back into its trap adapter.*

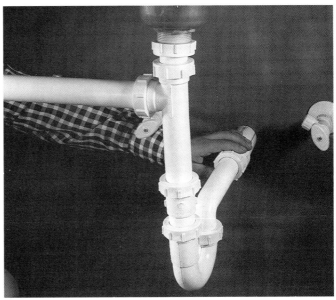

23 *The trap can be swiveled to reach several inches to one side to connect an off-center waste opening.*

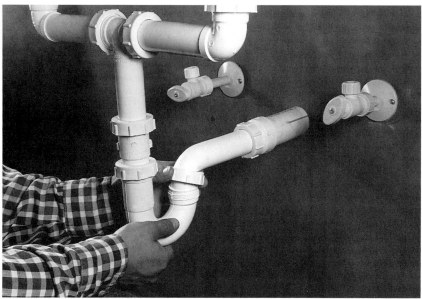

24 *To install a trap, get both parts into position before making the center joint. Don't tighten any parts until all are in place.*

186101 — Fig. 13) is used when the waste opening is centered between the two bowls.

Baffle Tee

A fitting that is always part of an end-opening continuous waste package is a baffle tee. It goes beneath the bowl on the waste-connection side. Genova's Part No. 186301 baffle tee contains an inner polypropylene baffle that keeps food disposer wastes from backing up in the sink drain. The baffle can be easily removed if you will not be installing a food waste disposer (Fig. 14).

Adjustable Tailpiece

Straight sink drain tubes come in two forms: adjustable tailpieces and flanged tailpieces.

Adjustable tailpieces contain slip fittings and are used to extend drainage tubes. If your trap is a little short of reaching up to the sink, an adjustable tailpiece is what you need. (See Fig. 15). They're also needed for connecting the drainage tubes to some food waste disposers.

Flanged Tailpiece

Flanged tailpieces, on the other hand, are simple tubes with a flat face on one end. They're used to connect the drainage tubes directly to a sink's basket strainers (and to some food waste disposers).

Both flanged tailpieces and adjustable tailpieces are 1½ in. in diameter and come in 6 in. and 12 in. lengths.

Flanged Waste Tailpiece

A short tube called a flanged waste tailpiece is used to connect a continuous waste to the sink drains. Two of these come as part of a continuous waste package.

As with most plumbing products, it's best to buy all parts from the same manufacturer. That way, you know that everything is compatible. Besides these fittings, you can buy extra slip jam nuts (Genova Part No. 147151) and slip washers (Genova Part No. 148001). A package contains assorted types and sizes, including flat washers for use with basket strainers.

Connecting Sink Drains

When plumbing the drains for a sink, always start at the sink's basket strainers. Basket strainers are threaded on the bottom of the sink to accept slip jam nuts. The nuts go onto 1½ in. flanged waste tailpieces and hold them to the strainers with flat washers between (Fig. 16). Some continuous waste packages, like Genova's, include everything needed to connect the sink. Others omit the waste tailpieces, so you must buy them separately.

When fitting the continuous waste for a double-bowl sink, the horizontal tube (or tubes) that connects the two bowls is usually long enough to suit standard sinks. If your sink's bowls are closer together than the standard, you'll have to trim the tube(s) to fit (Fig. 17) with a fine-tooth saw. Then fit up all the parts, starting with the baffle tee. Fig. 19 shows fitting an end-opening continuous waste. Be sure to get the baffle tee on the proper side (Fig. 20) to connect to the sink's trap. Fig. 21 shows fitting a center-opening con-

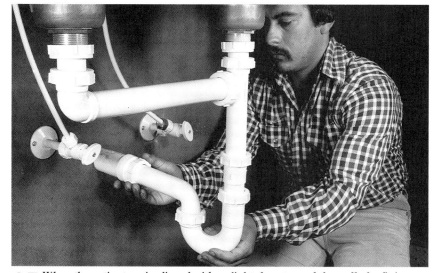

25 *When the entire trap is aligned with a slight slope toward the wall, the fittings may be hand tightened.*

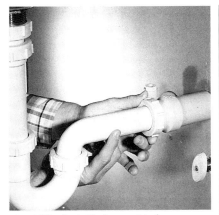

26 *The last slip jam nut to be tightened is the one on the trap adapter. If there are leaks when the bowl is test-drained, tighten those nuts a little more with water pump pliers.*

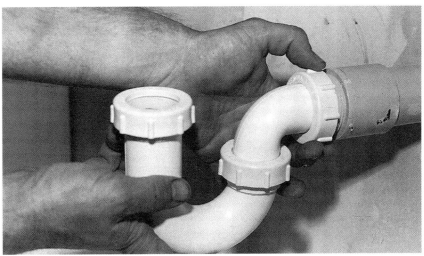

27 *A tubular P-trap (Genova Part No. 175151) fastens into the DWV system with a trap adapter and a slip jam nut.*

tinuous waste that has no baffle tee and does not need one. The hookups for an end-opening or a center-opening continuous waste are similar.

Waste Disposers

If your sink has been plumbed with two separate waste openings, you won't need a continuous waste setup. Instead, you'll need two flanged tailpieces and a pair of P-traps—one for each bowl. Double waste openings are intended to make it easy to hook up a food waste disposer. Still, it isn't difficult with just one waste opening, as long as a baffle tee is used to aim disposer discharges toward the trap. Speaking of that, if a food waste disposer is to be added along with new tubular drains, install the disposer before you connect the drains.

Connecting the Trap

The baffle tee and tailpieces should be long enough to connect the trap, leaving its arm inclined slightly in the direction of the water flow. If the tailpieces are too long, they may be cut to fit. Traps can be swiveled one way or the other to reach an off-center waste opening (Fig. 22 and 23). When working with an S-trap, make your cuts so that the trap's bend is up high and out of the way of any cabinet shelving.

Here's the proper way to install a trap. First, slide a slip jam nut (threads facing downward) up onto

the fixture tailpiece or the baffle tee. Then add a slip washer, with its flat side up (a slip washer's flat face should always aim toward the slip jam nut). The washer will grip the tailpiece, holding the nut in place. Now, push the trap's J-bend up to meet the nut, and run the threads up just tight enough to keep everything from slipping off. Slide another slip jam nut, threads first, onto the trap's arm, followed by still another nut, threads last. Slide a slip washer onto the trap arm, flat face first. If you can, use the pro plumber's leak-prevention trick shown in Fig. 7.

Now, slide the trap's arm into the trap adapter at its waste opening. If the arm doesn't quite reach, it may be extended with an adjustable tailpiece. If it's too long, you can cut it shorter. Align the two parts of the trap so they meet at the center swivel joint, and start the threads (Fig. 24). Then thread the slip jam nut onto the trap adapter. Don't tighten anything yet, though.

At this point, you're ready to slide the trap's J-bend up or down on the fixture tailpiece or baffle tee to get the desired slight slope in the trap arm. Align everything while loose (Fig. 25), then tighten that slip nut. Then tighten the slip nut at the trap adapter (Fig. 26). Finally, tighten the slip jam nut at the center joint.

Where local codes don't permit slip coupled traps, you can use a sol-

28 *A Los Angeles pattern trap (Genova Part No. 78715) joins directly to the waste pipe by solvent welding. This special trap is designed to eliminate slip connections to a fixture's waste pipe. A few local codes require it.*

vent welded Los Angeles pattern P-trap. Figs. 27 and 28 show a comparison between a slip coupled tubular P-trap and a solvent welded trap.

Do-It-Yourself Plumbing is an excellent softcover book (159 pages) on home plumbing and includes many step-by-step photos. It shows how to work with plastic pipe, and how to install plumbing for every room in the house. Special chapters cover sprinkler systems, and home sewer and septic systems. To obtain this book send $11.45 (postage paid) to Genova Products, Inc., 7034 East Court St., Box 309, Davison, MI 48423-0309. — *by Richard Day.*

Update your Bathroom with Ceramic Tile Flooring

Ceramic tile is used in 70%
of all bath floor installations.
Here's how you can do it, too.

1 *Remove the old floor covering. An electric heat gun can help soften the adhesive for removal, but use heat very carefully to prevent fire.*

2 *After removing the old adhesive, measure and mark the center points on all four walls. Snap chalk lines at the center marks. The intersection of the two chalk lines is the center point of the floor. Start laying tile at this point.*

As you view your tired bathroom floor, you may be debating what to replace it with. Vinyl tile or sheet goods may seem a good bet for a do-it-yourself job. Certainly vinyl is easy for the homeowner to work with. But the latest survey by the National Kitchen and Bath Association (NKBA) reveals that ceramic tile is the preferred material choice for bath floors by a margin of 2 to 1. Ceramic tile is easy to clean, durable and attractive, which accounts for its popularity. But now, new materials such as premixed tile adhesives, grout and sealers help make ceramic tile a good do-it-yourself project as well.

Select the tile color and pattern with the help of your tile dealer. Then measure the floor carefully to be sure you have bought enough tile to complete the job, with leftover tiles to save for future repairs.

The next step is to assemble the necessary tools: one secret of the pros is to never try to "make do," but to have on hand the tools that will make the job as easy as possible. Tools for ceramic tile work can usually be rented or borrowed from the tile dealer. Depending on the type of tile you buy, you may need a tile cutter, nippers for trimming tile corners or edges, a notched trowel for spreading tile mastic, a rubber squeegee, sponge and scrub pail, a

3 *Use a notched trowel or spreading tool to spread the tile adhesive. Note the spread rate instructions on the can label. If the adhesive is applied too thickly, it will ooze up between the tiles.*

carpenter's level, a chalk line, a measuring tape, a pencil, rubber gloves and a mixing stick for mixing powder grout. Also, a tungsten carbide blade and saber saw can be very helpful for making curved or scroll cuts. A kneeling pad such as those found in garden stores can prevent your knees from becoming sore as you kneel on the bare floor.

A flat, long-handled floor scraper can take some of the back-breaking work out of removing old flooring, or you can use a handheld scraper to lift old vinyl tile. An electric heat gun can be used to soften old adhesive and make vinyl tile or sheet vinyl removal easier. Depending on your floor level, it may be easier to lay new exterior-grade plywood over the old floor covering and start with a new substrate for your new tile floor. If you remove the old vinyl, ask your dealer for a solvent to remove the residual adhesive.

4 *Position the tile sheets carefully, checking the edges frequently to be sure the joints are straight. Pat the tiles in place with the palm of your hand.*

5 *Use a clean cloth, dipped in solvent, to remove any adhesive from the face of the tile. The adhesive will wipe away easily as long as it has not been allowed to set.*

6 *Continue laying the tile sheets, following the chalk lines and checking often to be sure the joints are straight.*

When you are ready to begin laying down the ceramic tile, plan your layout carefully. The borders should be even on all sides. This means,that rather than having a full tile at one wall and a half tile at the other, you should have a row of three-quarter tiles around the perimeter of the floor so it looks balanced and even.

To insure a professional-looking layout, measure the length of the room and mark the midpoint on both walls.

Then measure and mark the midpoint of the width of the room. Snap a chalk line in both directions. Lay

7 *Use tile nippers to shape the edges and corners of the tiles. The trick to using tile nippers is to chip away small bites at a time — don't try to break off large sections with one bite.*

your first tile(s) at the corners where the two chalk lines intersect.

As you continue laying the tiles, check frequently to be sure that the rows are straight. When all the floor tiles are laid, place a barrier in the bathroom door so people will not walk on the new tile until the adhesive is set. Let the tile adhesive set as recommended by the manufacturer, usually overnight or 24 hours, before applying the tile grout.

The grout seals the joints between the tiles so moisture cannot enter. You may choose white grout, or a grout in a contrasting color for accent. — *by Gary Branson.*

8 *Apply grout after the adhesive has set, usually after 24 hours. Use premixed grout, or mix grout in a plastic pail using a paint stir stick.*

9 *Use a floor sponge and clean water to wash away the grout haze from the tile surface. Change the water frequently so you are washing with clean water.*

10 *A disposable sponge brush is a good tool for spreading the silicone sealer. The sealer protects the grout from stains and moisture damage.*

Install a Countertop

A low-cost facelift that adds value to your kitchen.

1 *Determine the sink location, turn the sink upside down and trace the perimeter with a pencil.*

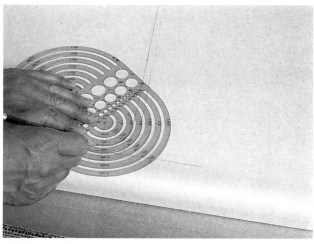

2 *Draw an inner perimeter for the actual cutout, using a straightedge and circle template (or compass). This inner dimension is specified in the sink installation instructions.*

If chips, burns and stains mar your countertop, then consider replacing it with a new one. Installing a new countertop is not difficult with proper planning.

Purchase the countertop at a home center store or lumberyard. If possible, let the retailer cut the countertop to the proper length, and apply the left and right end pieces which cover the exposed countertop edges.

Carefully measure the sink location to match that of the present sink. If a new sink will also be installed,

as in the project photographed here, measure the width, length and depth, then mark its location. Also, inspect the new countertop to insure the sink will clear the front of the cabinet.

Then turn the sink upside down on top of the new countertop at the desired location and draw the sink perimeter with a pencil. Use a straightedge and compass to mark the actual cutout which is usually about 3/16 in. inside the perimeter that is drawn on each side. The manufacturer of the new sink usually provides information on the size of

The old countertop was well-worn and in need of an update.

this cutout. If the old sink will be reinstalled, make sure that the 3/16 in. dimension is suitable.

Drill a starter hole in the countertop with a 1/2 in. twist drill; cut the waste away with a saber saw equipped with a fine-tooth blade. To eliminate scratching the countertop with the saber saw, apply several strips of tape to the bottom of the saw pad.

It is usually possible to cut only the left, right and front sides of the waste area with the good side of the counter up. To complete the cut, turn the counter upside down and finish the cut from the underside. To draw this new line on the underside, drill two small pilot holes tangent to, and inside of, the waste area from the laminate side; align the straightedge with the drilled holes in order to strike a new line.

Sources:

Countertop: Ralph Wilson Plastics Company, 600 General Bruce Drive, Temple, TX 76504. (800) 433-3222, (800) 792-6000 in Texas.

Faucet (F42B21944), Stainless Steel Sink (42B8903), Table and Chairs (1B2545N), Kenmore Dishwasher (22AP14795N): Sears Catalog.

Vinyl flooring: Century Solarian (24612) Armstrong World Industries, Inc., P.O. Box 3001, Lancaster, PA 17604.

3 *Drill a starter hole into the waste area and cut it with a saber saw equipped with a fine-tooth blade.*

4 *The back edge of the waste area will probably have to be cut from the underside of the countertop. Locate this cutting line by drilling pilot holes inside the waste area and strike a new line. Then use a saber saw to cut the back edge.*

Make sure that the waste area is firmly held while making the back cut to avoid tearing the laminate.

Cut Laminate to Length

If it is necessary to cut the laminate to length, make sure that the circular saw is equipped with a fine-tooth, carbide-tipped blade and that the blade is perpendicular. Draw the cutting line with the aid of a carpenter's square, and carefully cut the backsplash area of the countertop first. It is best to use a straightedge firmly secured to the backsplash.

Position the counter so that the underside is on top and, again, firmly position a straightedge for cutting.

Then make the finish cut while firmly holding onto the waste area.

Apply Laminate Edges

Next, custom cut, glue and secure filler strips to the countertop edges to accommodate the laminate edging.

The laminate edging is applied with an adhesive or secured with an iron. If adhesive is used, apply it to both the counter edge and the edging. Allow it to dry thoroughly, then carefully align the edging before contacting it with the countertop. Carefully rub the edging with a J-roller. If this specialized tool is not available, place a scrap piece of wood on the center of the

laminate edging, and tap the wood lightly with a hammer.

Edging with heat-sensitive adhesive requires using an iron to bond the adhesive to the countertop edge. Follow the manufacturer's instructions for applying this edging.

Now trim the edging with a laminate trimming bit installed in a router. Maintain firm control of the router, and position it so it sits flat on the edge or the laminate may be gouged. Remember to wear safety goggles when routing.

Attach Sink and Hardware

Install the new faucet and drain fittings to the sink, using plumber's putty to help seal the joints.

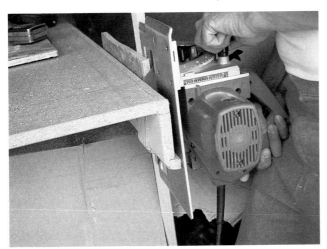

5 *Cut the countertop by first cutting the back of the backsplash. Use a straightedge guide and a temporary filler strip to insure a straight, square cut.*

6 *Attach a new straightedge and temporary filler strips to complete the cutting of the countertop. Then run the circular saw along this edge.*

7 *Custom cut, glue and secure filler strips to both edges of the countertop as shown.*

8 *After applying the laminate edging, rout the excess laminate with a flushing bit. Work carefully to avoid gouging.*

9 *Install the drain hardware. Apply plumber's putty to the area to be sealed to insure a leakproof joint.*

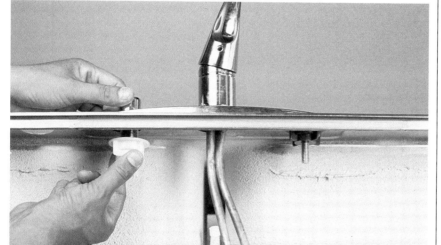

10 *Securing the faucet to the sink now eliminates an awkward installation from inside the cabinets after the countertop is in place.*

11 *Complete installation by connecting the sink drains to the kitchen plumbing.*

Sink fasteners are supplied with the sink. Insert the sink fasteners onto the sink and place the sink in the countertop's cutout. Secure the sink to the countertop by tightening the fasteners.

Remove Old Countertop

Begin by running a utility knife along the caulked areas of the countertop. Then turn off the water supply and any electrical lines that connect to a garbage disposal.

The countertop is held in place with screws driven from underneath the kitchen cabinets. Remove the drawers and goods from kitchen cabinets to access and remove these fasteners.

Disconnect the water lines from the existing faucets and the plumbing from the sink. If there is a garbage disposal, remove it.

With a helper, carefully lift the countertop and remove it, watching the caulked areas to make sure that the drywall or wallcovering does not pull away. If either loosens, run the utility knife once more along the caulked areas, then try again to remove the countertop.

Install New Countertop

Again with a helper, position the new countertop on the kitchen cabinets. Check for alignment, and then secure the countertop to the kitchen cabinets by reinstalling screws. Then connect the water supply lines to the faucet. Attach the garbage disposal and plumbing to the sink. Depending on the sink selected, it may be necessary to modify the plumbing to suit a deeper or shallower basin.

Complete the project by applying an exterior-grade caulk along the top of the backsplash to suit your decor. — *by Jonathan Wesley.*

Classic Shaker Table

**Shaker design is still as functional
today as it was over 100 years ago.**

1 *Glue up four boards for the top. Alternate the direction of the board end rings, and clamp on both sides. Make sure that the assembly is flat.*

2 *Cut the tenons with a tenoning jig or back saw.*

S haker purity in furniture design was being developed during the same period the "outside world" was embracing Victorian styles. In mid-19th century, Mother Ann, originator of the Shaker movement, directed her craftsmen to simplify their furniture designs. The Shakers believed functionality was its own beauty. Coincidentally, it was Queen Victoria that made ornate adornment fashionable.

Divergence between the two styles grew more extreme. Shaker curtainless windows contrasted with heavy, Victorian velvet draperies. The sect's small braided rugs on bare floors differed markedly from room-size Axminster carpets. White Shaker walls compared starkly with the bold, floral pattern wallpaper of the outside world. At the end of the 19th century, completely unadorned Shaker furniture contrasted sharply with the richly carved, rococo mahogany and rosewood furniture Victorians chose.

Ultimately, Victorian ornateness ran its course. After the turn of the century, the trend was away from heavy Victoriana. On the other hand, Shaker designs, along with simple Chinese designs, inspired modern

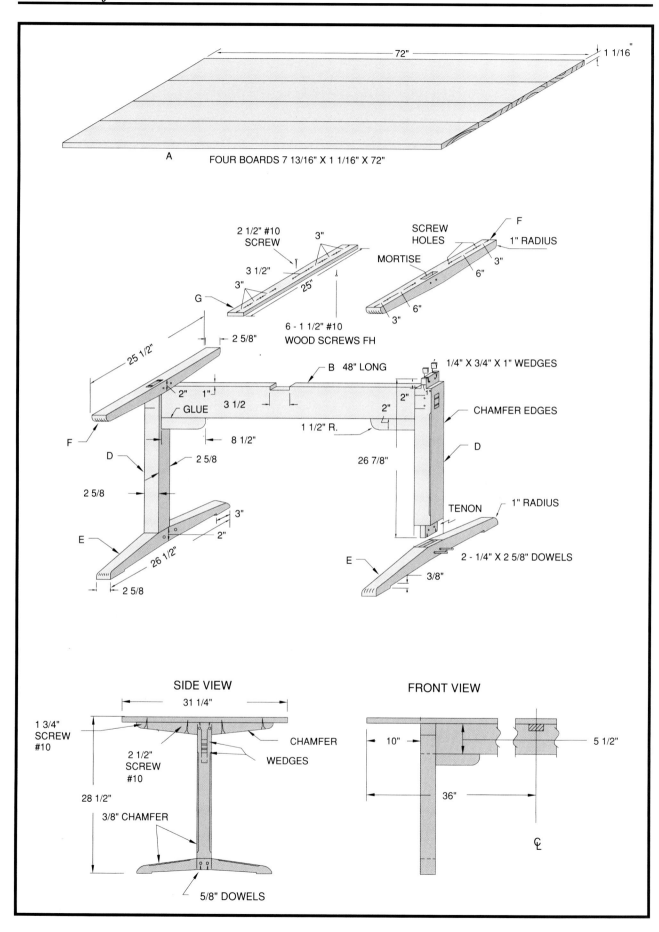

72"

1 1/16"

A   FOUR BOARDS 7 13/16" X 1 1/16" X 72"

2 1/2" #10 SCREW   3"

3 1/2"   3"

3"   25"

G

6 - 1 1/2" #10 WOOD SCREWS FH

SCREW HOLES   F   1" RADIUS

MORTISE   3"

6"

6"

3"

25 1/2"   2 5/8"

B   48" LONG   1/4" X 3/4" X 1" WEDGES

2"   1"   2"

GLUE   3 1/2   2"   CHAMFER EDGES

F   1 1/2" R.

8 1/2"   26 7/8"   D

D   2 5/8

2 5/8   TENON   1" RADIUS

E   3"   2 - 1/4" X 2 5/8" DOWELS

26 1/2"   E   3/8"

2 5/8

SIDE VIEW   FRONT VIEW

31 1/4"

1 3/4" SCREW #10   CHAMFER   10"   5 1/2"

2 1/2" SCREW #10   WEDGES

28 1/2"   36"

3/8" CHAMFER   ₵L

5/8" DOWELS

3 *Cut 45 degree stopped chamfers with a block plane. Mark limits with a pencil. Round and clean the chamfer ends by paring with a sharp 3/4 in. chisel.*

4 *Round the ends of the feet and top supports with a band saw, or use a chairmaker's saw as the Shakers did.*

furniture designers throughout the western world. While the Shaker religious sect has virtually disappeared, their furniture designs grow more popular.

This trestle table design appeared in many of their communities. Only the length varied. The table is elegant in its simplicity, functional in its design, appropriate in its material and straightforward in its construction. It is a classic you can build to beautify your home.

We use the low-back, two-slat chairs Shakers designed to store under the table. Chair tapes can be selected to coordinate with a color scheme, then changed later if the color scheme changes.

Shakers used cherry or maple for the trestle. A pine top, screwed to the trestle, was used for inexpensive replacement. Although a pine trestle would shave costs, Shaker joinery design requires a hard wood so the joints won't enlarge and make the trestle wobbly. We chose cherry because of its workability and beauty. The top and center stretcher dimensions may be lengthened to accommodate more seating.

Top

Start table construction by cutting and surfacing 5/4 pine to 1 1/16 in. thick. Joint the edges to be glued. Be sure the ends of the joints are tight.

Then clamp so the top is flat, not bowed. Check with a straightedge. Scrape and wipe off the excess glue. When it is dry, carefully plane and sand the top flat with 120 grit paper, and cut to 72 in. long. Shave the ends smooth with a block plane. Round all edges with sandpaper.

Trestle

Next, surface the trestle parts to size using jack, jointer and smoothing planes. Cut the tenons with slight (1/16 in.) taper to accommodate wedges. Cut through the mortises to fit tightly. Number each mating joint.

Round the feet and top support pieces, using a chairmaker's saw to cut the 1 in. radius. Use a rip saw to cut the lower tapers in the feet pieces. Save the cutoffs for wedges. Clean with sandpaper. Cut stopped chamfers in the uprights, feet and supports with a chisel, block plane or spokeshave.

Next, cut the center stretcher piece to size. Cut the tenons per the drawing. Then mortise the uprights to fit. Cut a 1 1/2 in. radius on two end pieces, and glue these decorative ends onto the center stretcher. Cut a notch in the stretcher for the center cross-support. Do a preliminary

BILL OF MATERIALS — Classic Shaker Table

Finished Dimensions in Inches

A	Top	1 1/16 x 7 13/16 x 72 pine	4
B	Trestle Center Stretcher	52 1/8 x 7 1/2 x 1 1/16 cherry	1
C	Trestle Center Stretcher	1 1/16 x 2 x 8 1/2 cherry	2
D	Upright	2 5/8 x 2 5/8 x 27 cherry	2
E	Feet	2 5/8 x 2 5/8 x 26 5/8 cherry	2
F	Support	2 x 2 5/8 x 25 1/2 cherry	2
G	Center Crosspiece	3/4 x 3 1/2 x 25 cherry	1

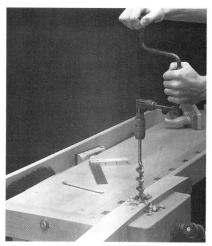

5 *Bore holes for the mortise, then cut square with chisels.*

6 *Clamp the trestle to dry. Add wedges later. Cut a slot for the wedge with a dovetail saw.*

7 *Clamp the chair kit pieces with band clamps and square up with a bar clamp, making sure the chair sits square with the floor and flat on it. The seat is square when diagonal measurements are the same.*

8 *Put on warp with a foam pad between. Weave tape over and under to create a checkerboard pattern.*

and wedges. Assemble and clamp. Check the trestle for squareness. When it is dry, bore ¼ in. holes for the pins. Put the pins in with glue. When dry, finish sand the entire trestle with 120 grit paper on a sanding block. Do not round the chamfer edges.

Screw the center crosspiece to the trestle. Flip the top upside down, placing the trestle on it. Mark and bore ³⁄₁₆ in. holes in the cherry cross-support and ⁷⁄₆₄ in. pilot holes in the pine top for No. 10 by 1¾ in. flathead wood screws. Countersink holes in the top supports. Screw the trestle to the top per the drawing.

Finishing

Rub cherry stain on top. Let it soak in for 10 minutes, then wipe and let dry overnight. Finish the entire table with Watco oil. Do the underside first, then flip the table over and oil the top and trestle. Apply three or more coats. Allow each to dry overnight. Then apply paste wax. Your classic table is now ready for years of dining.

Full-size plans are available for $9.95 from Creative Features, Shaker Table, Suite 1317, Six North Michigan Ave., Chicago, IL 60602. Chair kits are available from Shaker Workshops, P.O. Box 1028, Concord, MA 01742-1028. Obtain cherry or maple from Craftsman Wood Service, 1735 West Cortland Court, Addison, IL 60101. Catalog $2. — *by David A. Warren. Color photo styling by Virginia R. Howley.*

sanding of all trestle parts with 80 grit sandpaper.

Dry-assemble for fit. Then glue a foot, upright and top cross-support together, making certain it is square. Apply plastic resin glue and drive in wedges. Clamp while drying. When dry, pare the wedge ends flush with a sharp chisel and sand flush. Then assemble the other side in the same way.

Dry-assemble the center stretcher board to two end legs. When you are satisfied with the assembly, apply glue to the mortise

TV TRAYS

Pretty, yet practical, these TV trays stack up neatly on a stand for easy storage.

When I designed these TV trays, I had several objectives in mind. The trays had to be functional, well proportioned and sturdy. Since we would be using them in the family room, I also wanted trays that would be easy to store and visually interesting.

I selected white oak because I wanted a light-colored wood. Birch, beech, maple — all would do well, but I like the grain and color of oak. Cherry and walnut add some color and visual interest to the trays. I used oak for the stand, and walnut for the rails and the arms that support the trays. Exposed joinery adds further detail.

The stand is designed to set against the wall with all four trays placed on one side. (If you intend to move the stand and trays around, you may want to place two trays on each side of the stand and redesign the feet for support.) The top of the stand is curved to complement the curve in the top and bottom edge of the trays.

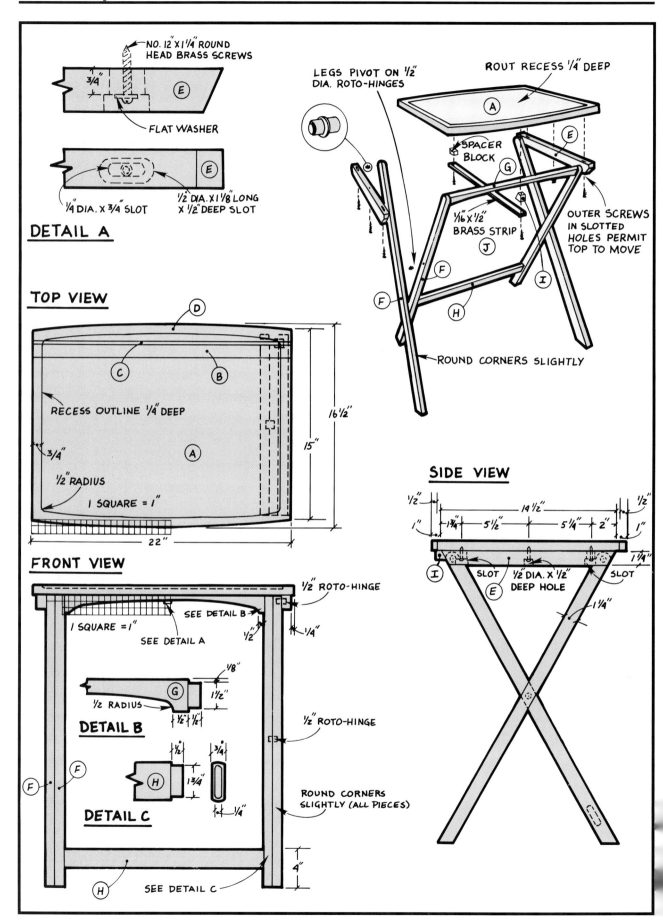

NO. 12" X 1¼" ROUND HEAD BRASS SCREWS

¾"

E

FLAT WASHER

E

¼" DIA. X ¾" SLOT

½ DIA. X 1⅛" LONG X ½" DEEP SLOT

DETAIL A

TOP VIEW

D

C

B

RECESS OUTLINE ¼" DEEP

¾"

A

½" RADIUS

1 SQUARE = 1"

22"

16½"

15"

FRONT VIEW

½" ROTO-HINGE

SEE DETAIL B

1 SQUARE = 1"

SEE DETAIL A

½"

¼"

⅛"

G

½ RADIUS

1½"

½" 1½"

DETAIL B

½" ROTO-HINGE

½"

¾"

H

1¾"

¼"

DETAIL C

F

F

ROUND CORNERS SLIGHTLY (ALL PIECES)

H

SEE DETAIL C

4"

LEGS PIVOT ON ½" DIA. ROTO-HINGES

ROUT RECESS ¼" DEEP

A

SPACER BLOCK

G

E

1/16 X ½" BRASS STRIP

J

F

F

OUTER SCREWS IN SLOTTED HOLES PERMIT TOP TO MOVE

I

H

ROUND CORNERS SLIGHTLY

SIDE VIEW

½"

½"

1"

1¾"

5½"

14½"

5¼"

2"

1"

I

SLOT

E

½ DIA. X ½" DEEP HOLE

SLOT

1¼"

1¼"

84

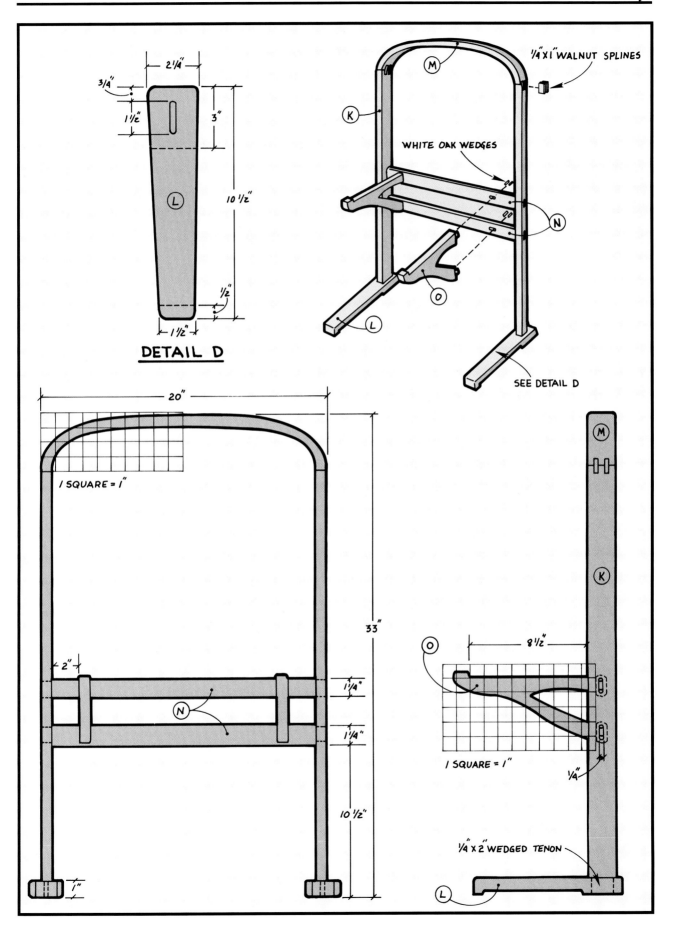

2¼"

3/4"

1½"

3"

10½"

½"

1½"

DETAIL D

¼"X1" WALNUT SPLINES

M

K

WHITE OAK WEDGES

N

O

L

SEE DETAIL D

20"

1 SQUARE = 1"

33"

2"

1¼"

N

1¼"

10½"

1"

M

K

O

8½"

1 SQUARE = 1"

¼"

¼" X 2" WEDGED TENON

L

A ¼ in. deep recess is routed in the top to prevent silverware and dishes from sliding off when carrying the tray.

Since the white oak section of the tray top (A) is 10½ in. wide, you'll probably have to glue this up from two 5 in. boards. Rip the pieces wide and joint or hand plane to final size. Rip and joint two 1 in. wide cherry strips (B), two ¼ in. wide white oak strips (C) and finally two 1½ in. wide walnut strips (D). Crosscut to length, allowing about ½ in. extra. Dry-clamp the top together. The joints should pull up tightly with only light pressure. If they don't, hand plane or joint the edges until they fit together using light pressure.

Since the trays will be wiped frequently with a damp sponge or cloth, I used plastic resin glue for water resistance. This may not be necessary if you use a surface finish such as lacquer or polyurethane varnish; it is just an added precaution. After the top has dried, scrape the dried glue off and flatten and smooth the surface with a hand plane or a belt sander.

Enlarge the pattern for the top from the drawing and transfer it to the wood. Using a band saw or saber saw, cut the top to shape, staying just to the outside of the cutting line. Smooth down to the line with a spokeshave or rasp and file. Once you have one top cut to shape, you can use it as a pattern to rout the other three. Use a band saw or saber saw to cut the remaining three tops roughly to shape, making them about 1/16 to 1/8 in. oversize. Clamp the finished top or pattern to the bottom of a blank. Then trim each top to its final size using a router with a flush trimming cutter set to ride on the pattern. If you make light cuts, this will leave a surface that requires only a light sanding to finish.

The next order of business is to enlarge the pattern for the recess and transfer it to a template. Don't forget to allow for the width and the thickness of the template guide you are using. Attach the template to the top. Hot glue works well, since it holds the template securely enough, but you can easily break it loose and

1 *Rout the perimeter of the tray top that is to be recessed. Use a template guide that has been temporarily hot-glued to the wood.*

chisel off the remaining glue. Using a ½ in. core box bit, set ¼ in. deep, run a groove around the tray top and then remove the template.

Rout the recess using as large a cutter as your router will handle. Don't rout yourself into a corner; that is, as you approach the last few passes, leave a small strip of wood to support the router. You can chisel the strip out after you finish routing.

Take it slow and careful as you approach the sides. Any slip here and you will have to start all over.

The next task is to smooth the bottom of the recess. Sand it smooth with 150 grit paper followed by 220 grit paper.

When you set the tray tops aside to work on other pieces, stack them one on top of the other with ½ in. or ¾ in. strips of wood in between. This

BILL OF MATERIALS — TV Trays

(For one stand and one tray)

Finished Dimensions in Inches

A	Tray Top	¾ x 10½ x 22 white oak	1
B	Tray Top	¾ x 1 x 22 cherry	2
C	Tray Top	¾ x ¼ x 22 white oak	2
D	Tray Top	¾ x 1½ x 22 walnut	2
E	End	¾ x 1¼ x 14½ white oak	2
F	Leg	¾ x 1¼ x 27 white oak	4
G	Top Rail	¾ x 1¾ x 18 walnut	1
H	Bottom Rail	¾ x 2 x 18 walnut	1
I	Guide Block	¾ x 13/16 x ¾ white oak	1
J	Strap	1/16 x ½ x 15¾ brass	1
K	Leg	¾ x 1¾ x 31¼ white oak	2
L	Feet	1 x 1¾ x 10½ white oak	2
M	Top	1¾ x 2¼ x 20 white oak	1
N	Rail	¾ x 1¼ x 20 walnut	2
O	Arm Support	¾ x 5½ x 10½ walnut	2

3 *Hot-glue the stand top (M) to a strip of scrap material and cut the ends flush as shown. Use a sharp blade.*

4 *Also hot-glue a clamping support to the arm (O) and clamp the arm to the rails (N). Drive in wedges into the tenons and allow the glue to dry. Then remove the clamp support.*

allows air to circulate evenly and prevents the tray tops from bowing.

Rip the ends (E) 1¼ in. wide from white oak and crosscut to length. Since you are screwing the ends across the grain of the tray tops, you must allow for expansion and

contraction with humidity changes. Therefore, the two outside screws are placed in slots that will allow the tops to move. The center screw does not need to move and is therefore placed in a hole.

Cut ¼ in. slots in the ends and

drill a ¼ in. hole in the center completely through the end to accommodate the screw shank. Then cut a ½ in. slot or hole ½ in. deep to provide for the screw head and washer. It is important that all the ½ in. holes or slots are cut exactly

½ in. deep or slightly less. If cut deeper, the 1¼ in. screws could protrude through the top.

Rip the legs (F) 1¼ in. wide from white oak and crosscut to length. Trim the ends of the legs and round over the top edges. Rip the top rail (G) and the bottom rail (H) to width and crosscut to length. Don't forget to allow the extra ½ in. on each end for the tenons.

When cutting a mortise and tenon joint, I find it easier to cut the mortise first and then cut the tenon a hair oversize and trim to fit with a sharp chisel. Dry-fit the leg and rails together to assure the joints pull up tight. Enlarge the pattern for the top rail and transfer it to the wood. Then cut to shape. You can use the first rail as a pattern to cut the other three rails, similar to cutting the tray tops. Round over the edges of the rails with a ½ in. round over bit set for a shallow cut. The edges are not rounded over completely (see the drawing).

Lay out the hole locations for the Roto-Hinges, and drill ½ in. diameter holes ⁹/₁₆ in. deep in the legs and ends.

Glue the rails and inside legs together. Glue the Roto-Hinge to the inside legs first. Then glue the Roto-Hinges to the outside legs and ends.

Clamp the ends to the bottom of the tray top and locate the screw holes. Drill pilot holes ½ in. deep, being careful not to drill through the top. Finish the tray top before you screw the legs in place. Finish the tray tops with three coats of polyurethane varnish, sanding between coats. Be sure to finish both sides of the tops the same or the tops will bow as one side picks up more moisture than the other side.

Cut the guide blocks (I) and the brass straps (J) to size, screwing to the tray tops with No. 6 by 1¼ in. brass flathead screws.

The stand is the next order of business. Rip ¾ in. thick white oak 1¾ in. wide for the legs (K) and crosscut to length. Cut the feet (L) from white oak and join to the legs with a mortise and tenon joint. Cut the mortises in the feet before cutting the tapered sides. Round over the front and back ends of the feet

and cut the relief on the bottom side with a band saw. Remove the band saw marks with a file or sandpaper, then round over all edges with a ¼ in. round over bit with a pilot guide. Cut the tenons on the end of the legs to fit the mortises in the feet.

The curved top (M) can be made in one of two ways. You can laminate it as I did, or, more simply, use a band saw to shape the curve from 8/4 white oak. If you choose to laminate the top, start by making the form. Enlarge the pattern and transfer it to the form, which is made by gluing together three pieces of ¾ in. plywood or particleboard. Use your band saw to shape the form, staying just outside of the line. Then smooth up to the line using sandpaper or a file. Cut the curved backer blocks from the scrap left over from shaping the form.

The laminates for this tight a curve should be about ¹/₁₆ in. thick; however, the best method is to cut a laminate and try to bend it around the form. If it doesn't bend easily, plane it slightly thinner and try bending it again. Proceed in this manner until the strip bends around the form. It is a good idea to cut the laminates about ½ in. wider and 1 in. longer than necessary, since they have a tendency to slip when glue is applied and the clamps are pulled up tight.

Clamp the laminates to the form, without glue, and check to make sure that the laminates pull up tight without gaps. If you have gaps, fine-tune the backer blocks by sanding or filing until the laminates pull up tight. Add glue and allow to dry overnight.

Trim the laminates to final size. I joined the top to the legs using two walnut splines in each side. Cut the ½ in. by 1 in. groove in the top and legs on the table saw using a ½ in. dado blade.

Rip the rails (N) from ¾ in. walnut, 1¼ in. wide. Crosscut to length, allowing extra length for the tenons to completely extend through the legs. Cut the mortises in the legs and the rails. The mortises extend completely through the legs and rails. Cut the tenons on the rails and the legs to fit the mortises in the legs and the feet, respectively. Round over

the ends of the tenons using a file to match the mortises. Round over the edges of the rails similar to the rails on the tray legs.

Dry-fit together to make sure the joints pull up tight. Then cut the slots in the rail and leg tenons with a band saw. Slope the slots inward toward the center of the tenons slightly and stop-drill the slots with ⅛ in. holes. (The holes prevent cracks from starting when driving the wedges home.) Add glue to the rails and legs, and clamp in place while the glue is wet. Drive the white oak wedges home. Cut the walnut splines for the top and glue the top to the legs. Then glue the feet to the legs.

Enlarge the pattern for the arm supports (O) and transfer to ¾ in. walnut, cutting to shape with a band saw or saber saw. Leave a tab on each of the arms for clamping. Cut the tenons to fit the mortises in the rails. Round over the tenons and cut the slots similar to the rails. Add glue and clamp the arms to the rails. Then drive the wedges home. After the glue has dried, trim the tabs off with a saber saw or coping saw.

Sand the stand with 150 grit paper followed by 220 grit. I finished the stand and legs of the trays with two coats of Watco natural oil. The second coat was wet sanded with No. 600 wet/dry paper. — *by Dennis Watson.*

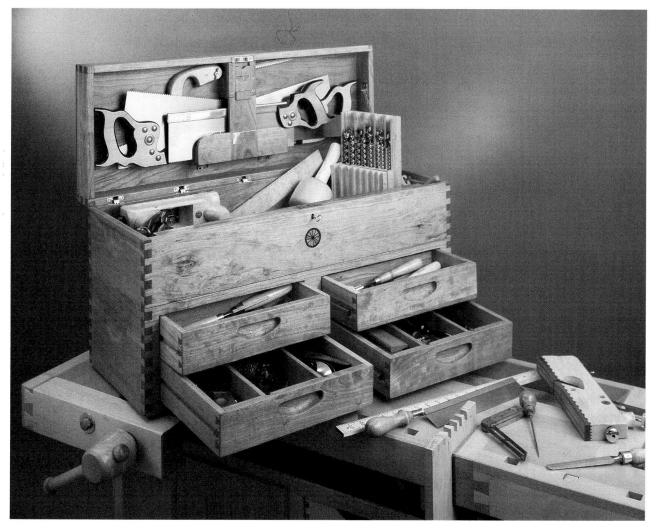

Through dovetails, sculpted handles, inlaid medallion and ebony initials are decorative touches that make this chest elegant.

Master Tool Chest

The original design has stood
the test of seven years of professional use.

While on an errand at the First National Bank in Chicago's loop, I happened to spot the original of this hand-dovetailed tool chest. I asked the union carpenter who owned it, "Can you take a minute to come down from your ladder so I can meet you?"

Patrick Murphy, a finish carpenter, slowly came down and cautiously listened as I introduced myself. I told him his master craftsman's tool chest had caught my eye and asked, "Can you take me on a quick tour of it?" Soon he relaxed, smiled and showed me several of the ideas he had incorporated in it. Some time later, we met in Pat's free time when he graciously explained more of the design details, and assisted

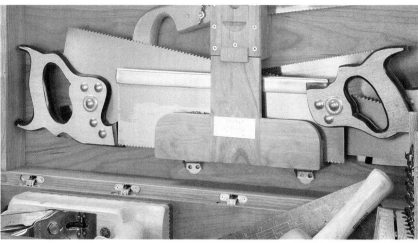

1 *Two halves bolt together. Aluminum angle inside bottom half gives support. Note tongue of framing square extending to lower corner.*

2 *Chest is sized to hold standard crosscut and rip saws and a framing square. All saws store in lid behind hinged spring-loaded holder.*

3 *Top unit nests into lower unit. Plywood bottoms on both units protrude ¹/4 in. to take any scraping action from dragging box, which saves the lower finished edges from splintering.*

4 *Space under lower drawers stores fold-over clipboard for notes and plans, sandpaper and screw-in handle for top unit. The area between drawers in lower unit provides space for retractable handle and drawer-locking mechanism.*

5 *Drawers lock when lid closes. Lid pushes pin down, forcing ¹/2 in. section of drawer slide down into mating mortise in drawer side.*

6 *Drawer backs have spring-loaded dowel pin stops so short drawers can't pull out accidentally and spill tools.*

7 *Dolly is made of oak with heavy-duty swivel casters so it may also be used as a tool for positioning large cabinets and cases. The larger the casters the more durable the dolly.*

8 *Furniture glides in the bottom of chest nest into recesses in dolly so tool box won't be jarred off dolly when moving.*

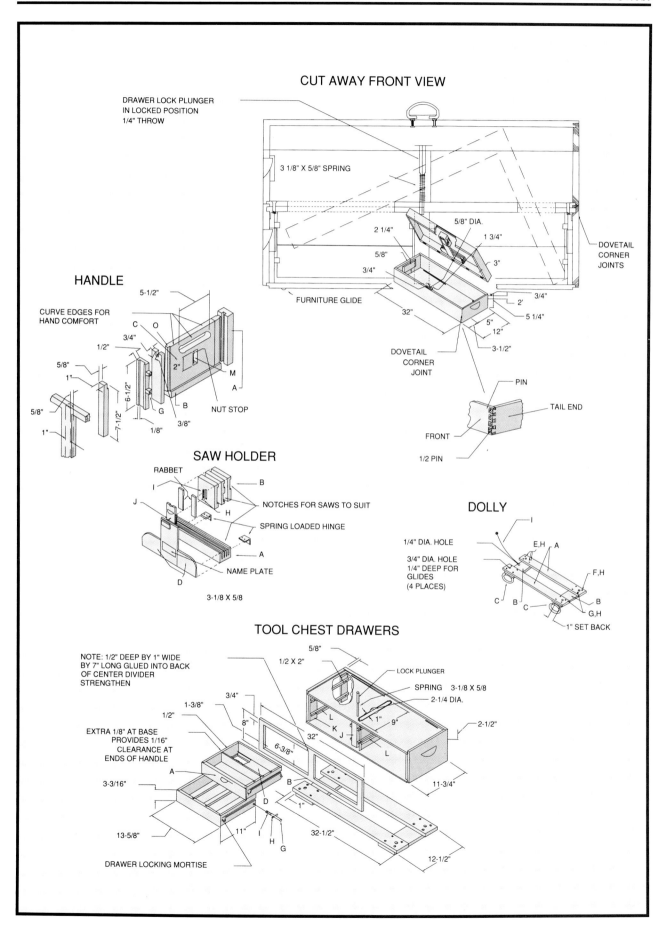

CUT AWAY FRONT VIEW

DRAWER LOCK PLUNGER IN LOCKED POSITION 1/4" THROW

3 1/8" X 5/8" SPRING

5/8" DIA.

2 1/4"

1 3/4"

5/8"

3"

3/4"

DOVETAIL CORNER JOINTS

FURNITURE GLIDE

32"

3/4"

2'

5 1/4"

5"

12"

3-1/2"

DOVETAIL CORNER JOINT

PIN

TAIL END

FRONT

1/2 PIN

HANDLE

CURVE EDGES FOR HAND COMFORT

5-1/2"

C   O

3/4"

1/2"

5/8"

1"

5/8"

1"

2"

6-1/2"

7-1/2"

M

A

B

G

NUT STOP

1/8"

3/8"

SAW HOLDER

RABBET

I

J

B

H

NOTCHES FOR SAWS TO SUIT

SPRING LOADED HINGE

A

NAME PLATE

D

3-1/8 X 5/8

DOLLY

1/4" DIA. HOLE

3/4" DIA. HOLE 1/4" DEEP FOR GLIDES (4 PLACES)

I

E,H   A

F,H

C

B

C

B

G,H

1" SET BACK

TOOL CHEST DRAWERS

NOTE: 1/2" DEEP BY 1" WIDE BY 7" LONG GLUED INTO BACK OF CENTER DIVIDER STRENGTHEN

5/8"

1/2 X 2"

LOCK PLUNGER

SPRING   3-1/8 X 5/8

2-1/4 DIA.

3/4"

1-3/8"

1/2"

8"

32"

6-3/8"

L

K J

1"

9"

2-1/2"

EXTRA 1/8" AT BASE PROVIDES 1/16" CLEARANCE AT ENDS OF HANDLE

A

B

L

3-3/16"

D

11-3/4"

I   H

1"

13-5/8"

11"

G

32-1/2"

12-1/2"

DRAWER LOCKING MORTISE

with the extensive measured drawings needed to reproduce his tool chest.

For centuries, artisans have displayed their skills in crafting their own tool boxes. These chests have traditionally been fitted with locking mechanisms to protect against theft. They were made of wood to absorb moisture and protect against rust.

Professional woodworkers still have to provide their own hand tools. Portable electric tools (excluding drill bits) are usually the responsibility of the employer. Because plans for wooden tool chests are scarce and the common gray steel tool box is widely available, woodworkers may forget that it is possible to custom-make a handsome chest. Such hand-crafted tool boxes will fit individual tool storage needs and provide a lifetime of satisfying use.

Patrick's tool chest design was developed over the years. It evolved as he talked to other woodworkers about their storage ideas. You can adapt Pat's well-planned tool chest features to your own needs. Or use these detailed measurements and instructions to duplicate Pat's chest. His design was adapted by Richard W. Eickhoff, maker of custom furniture and cabinets.

Your own wooden tool chest can be sized to store the hand tools you now own and also allow room for your collection to grow. It can reflect your artistic taste and be a proud statement of your skills as an artisan. Anyone who can build such a fine chest can also build any kind of case furniture.

First, a tool chest should be strong enough to carry the heavy weight of many tools. This design has survived the test of seven years of hard, daily use. Hand-cut dovetails are the strongest corner case joints. Pat says his box has been dragged through Chicago's streets, up and down curbs. It has withstood humidity extremes; the chest was left at construction sites for long weekends in bitter cold, very dry weather or in hot, humid conditions. It's a sturdy box.

Second, a tool chest should keep valuable tools safe. The cabinetmaker's chest is theft-

BILL OF MATERIALS — Master Tool Chest

Dolly

A	Long Rail	¾ x 3¾ x 32½ oak	2
B	Short Rail	¾ x 4¾ x 12½ oak	2
C	Caster	2 in. rubber wheel casters	4
D	Caster Lock	⅛ x 1 x 4 flat cold-rolled steel	2
E	Fastener	¼ x 2 F.H. mach. screw	12
F	Fastener	¼ x 2½ F.H. mach. screw	2
G	Fastener	¼ x 3 F.H. mach. screw	2
H	Fastener	¼ hex nuts	20
I	Rope	¼ x 36 nylon rope	1

Tool Chest

A	Lower Case Side	⅝ x 8 x 12 cherry*	2
B	Lower Case Front, Back	⅝ x 8 x 32 cherry*	2
C	Upper Case Side	⅝ x 5¼ x 12 cherry*	2
D	Upper Case Front, Back	⅝ x 5¼ x 32 cherry*	2
E	Lid Side	⅝ x 2¾ x 12 cherry*	2
F	Lid Front, Back	⅝ x 3 x 32 cherry*	2
G	Lid Top	⅝ x 12 x 32 cherry*	1
H	Side Lift Reinforcement	⅝ x 2½ x 6 cherry*	4
I	Plunger Guide	¾ x 1¾ x 5 cherry*	1
J	Front Spline	¾ x ⅝ x 7½ cherry*	1
K	Corner Brace	½ x 2 x 6¼ cherry*	4
L	Outer Drawer Runner	⁷⁄₁₆ x ½ x 10⅝ cherry*	4
M1	Bottom	½ x 11⅜ x 31⅞ fir plywood	1
M2	Case Panel	½ x 11⅜ x 31⅞ fir plywood	1
M3	Case Bottom	½ x 11⅜ x 31⅞ fir plywood	1
N	Hinge	No. 101 Soss hinges	5
O	Steel Rod	⅝ dia. x 6⅜	1
P	Spring	⅝ o.d. x 3⅛ spring	1
Q	Chest Lock	1½ x 2½ No. K34*	1
R	Handle	(for top of case)	1
S	Fastener	¼ x ¾ F.H. mach. screw	4
T	Fastener	¼ Teenut (to mount handle)	4
U	Aluminum Angle	⅛ x ¾ x ¾ x 10⅝	2
V	Fastener	¼ x 1¼ F.H. mach. screw	6
W	Fastener	No. 8 x ½ F.H. wood screw	8
X	Fastener	No. 6 x 1 F.H. wood screw	20
Y	Fastener	6d finishing nails	as required
Z	Bottom Furn. Glide	¾ dia.*	4

| AA | Hardware | Lid support chain, brackets | |
| | | and screws | as required |

Tool Chest Drawers

A	Front	⅝ x 3⅛ x 13⅝ cherry	4
B	Side	½ x 3⅛ x 10⅞ cherry	8
C	Back	½ x 3⅛ x 13⅝ cherry	4
D	Divider	¼ x 2¾ x 10⅜ cherry plywood*	4
E	Divider	¼ x 2¾ x 13⅛ cherry plywood*	4
F	Bottom	¼ x 10⅜ x 13⅛ cherry plywood*	4
G	Dowel Lock	⅜ dia. x 2½	8
H	Dowel Pin	¼ dia. x 1	8
I	Lock Spring	⅜ o.d. x 1	8

Sawholder

A	Base Retainer Block	1⅝ x 1⅞ x 10 cherry*	1
B	Retainer Block	⅝ x 3 x 3½ cherry*	3
C	Vertical Retainer	¼ x 3 x 6 cherry*	1
D	Horizontal Retainer	¼ x 2½ x 3½ cherry*	2
E	Catch Piece	¼ x 2 x 2¾ cherry*	1
F	Catch Piece Latch	¼ x ¾ x 2 cherry*	1
G	Catch Piece Retainer	¼ x ⅝ x 3¼ cherry*	2
H	Spring Dowel	¼ dia. x ⅜	1
I	Spring	¼ o.d. x 1½	1
J	Fastener	No. 5 x ½ F.H. wood screws	7
K	Hardware	Spring-loaded hinge*	2

Handle

A	Rear Female Guide	1⅛ x 1⅝ x 7 cherry	1
B	Base	⅞ x 1⅝ x 9⅛ cherry	1
C	Center Section	¾ x 6⅝ x 6¾ cherry	1
D	Male Guide	¾ x 1⅛ x 6⅝ cherry	2
E	Front Female Guide	¾ x 1⅝ x 7 cherry	1
F	Locking Mechanism	½ x 1⅝ x 6½ cherry	1
G	Runner Locks	⁷⁄₁₆ x ⁷⁄₁₆ x ½ cherry	4
H	Center Runner	⁷⁄₁₆ x ½ x 10⅛ cherry	4
I	Dowel	⅜ x 1⅛ birch	3
J	Compression Spring	¼ dia. x ½	2
K	Fastener	No. 6 x ¾ F.H. wood screws	12
L	Fastener	No. 8 x 1¼ F.H. wood screws	4
M	Fastener	¼ x 4¾ carriage bolt	1
N	Fastener	¼ hex nut	3
O	Fastener	¼ washer	2

Note: F.H. stands for flathead and mach. for machine

* Available from Craftsman Wood Service, 1735 West Cortland Ct., Addison, IL 60101.

proofed with concealed Soss hinges and a steel chest lock. Drawers open only by unlocking the lid and raising it so that the spring-loaded drawer locking dowel pin is released.

Besides being rugged and safe, this handsome chest has clever storage ideas. The division of space is excellent. The relatively small box holds important bulky tools, including a framing square, 28 in. long level and full-length crosscut and rip saws. It has room for a complete collection of planes, chisels, bits, marking and measuring tools, plus the myriad of hand tools that woodworkers collect.

Side lifts and drawer pulls are recessed; hinge knuckles are concealed. There is no metal on the exterior, so there is no chance of protruding hardware or metal corner reinforcements inadvertently gouging cabinetry at a work site. It also gives the chest a clean look.

The dolly was made of ¾ in. oak with simple overlap joints for strength. The heavy-duty dolly provides convenient "wheels" for the hefty tool chest, and is also a handy tool in itself. For example, the dolly can be used to move heavy cabinets. This design has oversize casters for that reason.

If all four casters swivel freely, the dolly and chest won't pull straight. On the other hand, when moving heavy cabinets in tight spots, there is an advantage if all casters swivel. In this design, the two rear casters lock to roll straight, but can be released to swivel. To lock them, a metal strap butts against the back of the caster shell. The metal strap is held in place by the bolt that holds the caster to the dolly. (See diagram.) More design features of the tool chest are illustrated in the photos.

This cabinetmaker's tool chest with dolly is not made quickly. It is a return to an age when life was lived at a slower pace and there was more time to indulge in the woodworking arts. Because it is custom-designed and takes time to craft, for most it is affordable only as a do-it-yourself project. But building such a fine piece is a labor of love.

Pat's chest was constructed of

oak. Rick selected cherry because of its beauty, ease of working and strength. Rick wanted his chest to have the natural look of contrasting light and dark cherry, so he did not match boards for grain or shade.

Construction

Before constructing the chest, it is helpful to draw a full-size section of the front and side views. This is especially true if you make any variations in these plans. A full-size section shows the precise position of slots for the framing square, drawer slides, etc.

Start construction by planing all cherry boards needed for the case sides and lid top to ⅝ in. thick. Cut the boards oversize. Butt join them with glue.

When dry, square the boards for the case sides and top to ⅟32 in. more than the finish length and width. Sand the surfaces to be placed inside thoroughly.

Dovetailing is one of the strongest ways to join ends of boards together at right angles. We used hand-cut through dovetails because they are the strongest and the most attractive for the tool chest. While hand-cut dovetails take time and patience, they are not as difficult nor as time consuming as they appear if you do each step methodically. Cutting a set of corner dovetails took Rick about one-half hour.

Dovetail Layout

There is room for variation in styling decorative through dovetails. Unlike machine-cut dovetails, hand-cut dovetails can be proportioned to any board width and made so they actually look handcrafted.

The strongest joint results when the pins and tails are equal in width. But for both appearance and strength, tails are commonly about one and one-half to two times the width of pins. An old standard is to make the widest part of the pin one-half the board thickness, or ⁵⁄16 in. for this case. Though not quite as strong, pins may be reduced to small, arrow-like wedges that were popular in 18th-century cabinetry. Another option is to vary the distance between pins by alternating one or two

narrow tails with a wide tail. Thus, two or three pins are grouped close together, followed by a wide tail.

The principal stress on the corner joints comes from heavy tools pressing against the long front and back boards, exerting mechanical strain to pull them away from the ends. Therefore, to take advantage of the flared pins and to counteract the mechanical strain, lay out the dovetail pins on the front and back boards. Lay out the tails on the end boards. (See diagram.)

The first step in laying out the pins is to set a marking gauge ⅟32 in. more than the ⅝ in. thickness of the boards, and to scribe a line around the ends of the front and back boards. After assembly, the pins and tails will protrude ⅟32 in. and can be sanded or planed flush for a tight, neat joint.

Lay out the center line for half pins at the bottom and top of each board. (They are called half pins because they are tapered only on the inner edge, not because they are half the width of full pins.)

Lay out the full pins by dividing the remaining space on the outer face of the board end into the number of pins to be cut. To divide this space equally, lay a rule obliquely on the face so that the number of inch graduations desired fits between the two half pin division marks.

To space out the dovetails for the upper section of the tool chest, center a pin on the line where the lid will be sawn free. Increase the width of this pin by the size of a saw kerf (about ³⁄16 in.) so that when the lid is sawn free, the remaining half pins will be the same size as the other half pins.

These divisions mark the center lines of the half and full pins. With a divider set at half the width of the pin, mark the width of the pins on either side of the center line. Use a square and sharp knife or awl to scribe these pin lines on the face of the board. Then set a bevel gauge for about 75 to 80 degrees, and mark the taper of the pins on the end grain. To avoid confusion, darken the waste areas with a soft lead pencil.

Before cutting, check the layout of pins on both ends of the board.

Make sure that the wide ends are along the inside surface, spaced correctly and of uniform size.

Cutting Pins

Hold the piece in a vise so that the tapered line being sawn is vertical, and cut the pins with a fine dovetail saw. The saw kerf should be on the waste side so it grazes the scribe mark. Do not saw into the marking gauge line.

Next, use a bench hold-down to clamp the piece horizontally on the bench. With the appropriate size wood chisel and mallet, deepen the marking gauge line to establish a definite edge along the mortise bottom.

Alternately remove chips and deepen the line at the bottom of the mortise. Cut successively heavier chips until half the waste is removed. Then flip the board so that the opposite marking gauge line is visible, and repeat the process on the other side.

Use a sharp chisel to shave the saw marks clean on each side of the pin. Cut the bottom of the socket at 90 degrees by paring along the marking gauge line so that the end-grain surface is dead flat, thus assuring a tight-fitting joint.

Marking Out Tails

Use the hand-cut pins as a pattern to scribe the tails on the corresponding end pieces. (Note the related boards at each corner.) Then place the long board with the pins at 90 degrees against the end board to be marked. Place the boards with the outer surfaces flush, the bottom of the pin sockets along the marking gauge line, both inside surfaces facing inward and with the widest ends of the pins toward the inside of the joint. Then scribe a mark around, and tight against, the angle of the pins. Use a square to scribe the lines across the end grain.

Shade the waste core with a soft lead pencil. Remove the waste between the tails with a saw and chisel.

Preliminary Assembly

When the waste has been removed and inside corners have been neatly pared, test-fit the joint. Do not force it, or it may chip or split. When correctly cut, the joint will fit snugly when it is tapped together. Dovetail all four corners of both halves in this manner.

Side Lifts

Glue a ⅝ in. thick reinforcement for the side lifts to the inside of each end piece. When dry, carve the four recessed lifts.

There are two ways to carve the lifts. Use a circular saw, making successive cuts and changing the blade depth with each new cut. Then clean with a gouge and sandpaper. Or, carve the lift out entirely with a gouge, which is safer and gives a handcrafted appearance. Use a gouge to undercut a trifle so the fingers have a concave area to grip.

Cut rabbets around the bottom edges of the upper and lower sections. Cut the dado around the upper edge of the bottom section.

Lower Case

Cut out the upper plywood panel to fit the dado. Bore a hole and saw out the retractable handle opening. Saw slots for the framing square. Because these slots weaken the panel, keep them as short as possible.

Saw out the two rectangular drawer openings in the front board of the lower section. Clean edges with a scraper and sandpaper held in a block.

For appearance, the face of the lower section was cut from one board so the grain runs one direction. This design reduces the center divider to short, weak end grain. Reinforce it with a spline down the center. This spline projects to serve as a vertical guide for the drawer locking mechanism.

Rout the stopped (or blind) dado in the back of the center divider. Cut and fit the matching spline. Glue it in place.

Assemble lower case sides and plywood top. Clamp together and check for squareness by verifying that both diagonal measurements are equal. Place scrap wood under the clamps to prevent them from crushing the wood.

Cutting the drawer openings in the front piece weakens the front

dovetail corner joint. However, the ½ in. by 2 in. corner braces reinforce these joints. The braces also provide fastening shims for the drawer runners because they extend beyond the lift reinforcement blocks. Spread glue liberally on the corner braces and screw in place with No. 6 by ¾ in. flathead wood screws.

Construct the retractable wooden handle components next. Cut the center section of the handle to size, and bore the extremities of the finger hole with a 1⅛ in. bit. Then cut out the waste between both holes with a jig saw.

To make the two male end guides, rabbet both surfaces of a piece about 16 in. long. Then cut the two end pieces 6⅛ in. long. Fasten each of these ends to the middle section with glue and three ⅜ in. dowel pins. When dry, round the handle edges smoothly with a plane, rasp and sandpaper.

Cut ⅜ in. by ¼ in. mating grooves in the front and back guides. Cut the base piece next. Counterbore two ¾ in. holes in the top and bottom faces of the base piece (or the diameter of the washer). Then bore a ¼ in. hole through the bottom of the handle piece into the stop nut recess.

Thread a carriage bolt up through the base, sinking the head and washer into the base. Slip a washer and screw a nut down into the base to secure the bolt.

Then slip a bolt up into the handle piece, and screw a double nut at the top to act as a stop. The handle should raise about 1⅝ inch.

Disassemble the handle components, and glue the rear female guide-piece to the back of the case.

The drawer locking mechanism Rick designed is an adaptation in wood of the classic cast-brass chest drawer lock. These brass locking systems were installed in beautiful 18th century cases made of fine woods and lined with silk. The chests were carried aboard ship on trips between the New and Old Worlds. In Rick's design, the springs are installed so they fit into top and bottom recesses. If they break from metal fatigue, they can be compressed, removed and replacements installed.

Make the movable piece and cut the center groove in the front. Bore ¼ in. diameter recesses in the bottom for the springs.

Cut out the plywood bottom panel to fit the rabbet joint in the bottom of the lower section. Then rabbet the edge of the plywood. Next, mark and bore ¼ in. diameter recesses in the plywood to accept the springs.

Next, invert the lower section, then insert the handle assembly and drawer locking guide. Glue the plywood bottom into its rabbet in the case bottom. Also, spread glue sparingly on the plywood where it touches the base of the handle. Use several 6d finishing nails to hold the plywood bottom tight in its rabbet while the glue dries. Put ¾ in. diameter furniture glides on the base to protect the plywood and allow it to fit into the mating holes in the dolly.

Turn the section upright, and weight the handle down so the glue joint at the base is tight against the plywood. Be sure the drawer locking mechanism moves freely. Install compression springs. When dry, invert the section and reinforce the glue joint of the handle base with three No. 8 by 1¼ in. flathead wood screws.

Drawers

Fasten the drawer runner to the sides of the cabinet with glue and No. 6 by 1 in. flathead wood screws. Then install the drawer slides on the middle sections, including the short movable pieces.

For a drawer to work smoothly, it must stay square, despite the strain put on it. If it twists when pulled, it will bind and stick. More than any other joint, dovetailing resists twisting a drawer out of square, which is why it is so common in drawer construction.

Start by cutting out four ⅝ in. thick boards for the front and twelve ½ in. thick boards for the four backs and the eight sides. Sand and mark inside surfaces. Lay out the half-blind dovetail for the fronts.

Set a marking gauge for ½ in., or the thickness of the drawer sides. Scribe a line on the back side of both ends of each of the four drawer fronts. Then scribe a ½ in. line in the end grain of each drawer front. Using the same setting, scribe a mark all around each drawer side.

For the half-blind dovetails, cut the tails first, then mark the pins. Lay out the tails on the drawer side pieces. Saw and chisel the core waste between tails in the same way the through dovetails were done.

When the tails have been cleaned, use them as a pattern to scribe the pins on the end grain of the drawer fronts. Hold the front in a vise, and place the cut pins over the end grain with the inside surfaces toward one another. Keep the board edges lined up and the ends of the tails along the marking gauge line. Hold the side securely as you scribe the pins on the end grain. Extend the lines down the inside of the drawer front with a square.

Cut pins by sawing the inside edge of the drawer front. Remember to graze the line on the waste side. Saw at an angle down to both marking gauge lines. Remove the waste by chiseling, in the same way the through dovetails were done. When the saw nears the bottom of the cut, extend the depth of cut in the sides with a chisel and mallet. Continue removing chips. Clean up corners carefully and test-fit the joint.

Cut the blind dovetail at the other end of the front piece. Then cut the through dovetail joints in the back to complete the four corners. Cut two grooves for drawer dividers in the sides, back and front, so the area is divided into three equal spaces. This division also permits flexible divider arrangements.

Carve the pulls in the drawer fronts to match the side lifts.

Next, cut the joint for the bottom. There are at least two ways of joining the bottom. Dado the sides and slide the bottom in after the sides have been assembled, or cut a rabbet in flush. A bottom dadoed in is a strong joint. A bottom rabbeted flush is easier to do.

The reason for a flush bottom is that these drawers will get filled to capacity, and every quarter inch is useful space. Furthermore, when drawers get full to the brim, a screwdriver or chisel pointing up cannot get caught on the flush bottom of the drawer above.

After assembly, fit the drawers and mark the location on the bottom of each. Cut and install dividers as desired.

Upper Carcass

Next, cut the half-blind tongue and rabbet joint for the lid top. Use a circular saw dado head, or router, to cut this joint around the edge of the cherry lid board. Then cut mating grooves in a scrap piece to test for size before cutting actual grooves in the upper section side and end pieces.

Dry-assemble the upper case. Then coat the side and top joints with glue. Assemble the long front and back pieces to the lid top first, then join the end pieces.

Gluing the top of the lid joints is easier if you have four hands. Clamp the assembly together and check carefully for squareness. When dry, saw the lid free from the upper section. The saw kerf should split the dovetails in half at each corner. Scribe the saw line, and watch the saw closely. Adjust it immediately if the case moves from the fence, which causes the blade to drift from the line.

A hollow-ground planer blade has the narrowest kerf and cuts edges the smoothest. If it is newly sharpened, the blade's tendency to burn the wood will be minimal.

After sawing the lid free, clean the edges of both case and lid, as necessary, with a smoothing plane and sandpaper.

Cut and fit the plywood panel bottom in the upper section. Cut slots for the framing square. Rabbet the edge so that it nests snugly into the top of the lower section. Glue in place.

Next, install the concealed hinges in the lid. The ⅝ in. stock requires five invisible hinges, due to the size of the top and the weight of the saws. Mark the location and outline of the hinges on the upper and lower edges. Bore the mortise holes, and clean waste with a chisel. Install the chain lid support so that the lid opens about 120 degrees.

Install the chest lock per the manufacturer's instructions. Rick chose a classic chest lock with key. A recessed combination lock designed for luggage may also be used. A combination lock eliminates the problem of a misplaced, forgotten or lost key.

Install the handle in the top next. To keep things neat, flush and strong, we adapted hex bolts ordinarily used for exterior door hinges. The adaptation provides two threaded tubes through the top in which to quickly bolt the handle.

To achieve this, cut the hex bolts off to ⅝ in. long. Drill a hole for tapping through the bolt head. Drill this hole through the threaded portion, but without touching the threads, so that it will act as a centering guide. We used a No. 6 drill and continued existing threads with a ¼ in. tap.

Bore a ⅝ in. hole ⅛ in. deep to countersink the bolt head into the top of the lid. Bore a 13/32 in. hole in the center of the ⅝ in. hole to accommodate the body of the bolt. Cut the screws to the proper length so each has an equal number of threads as it comes from the top and bottom, about 5/16 in. of threads.

Next, design the saw holder to accommodate the saw collection you have or plan to buy. Rick's tools included crosscut, rip, back and coping saws in the swing down piece. Behind it, in the block, are slots to hold the keyhole, tenon and veneer saws. The width and depth of the slots are cut to fit each saw. The length and curved corners of the spring-loaded saw holder are designed to just clear the framing square when the lid is swung down.

Saw the base piece of the saw holder to size, and cut the slots for your large saws. Cut the rabbeted spring-loaded catch piece, and rout the spring socket. Cut the vertical and horizontal pieces of ¼ in. cherry and glue them in place. Next, cut the three blocks for the saws, cutting the slots to fit your saws. Glue the three blocks together. When dry, fasten the back to the underside with glue. Fasten the base piece with spring-loaded hinges.

Rick's drawer-locking mechanism starts with the upper case. Cut the plunge guide, and bore the ⅝ in. hole. Turn the plunger on the lathe next. Or make the plunger by inserting a ⅜ in. dowel into the center of a ⅝ in. dowel. Fasten the two case sections together, mark the location for the ⅜ in. hole and bore it through the two cases. Then glue the plunger guide to the inside of the front board so that the plunger pin is aligned with the holes. Clamp and let dry in that position.

The saw holder pushes the drawer-locking mechanism plunger down. Complete installation of the drawer-locking mechanism next. Place the upper section on the lower section. Insert the plunger, and cut it so that when the lid is closed, the saw holder will depress the plunger ⅜ inch.

Aluminum Angle Reinforcement

The two cabinets are secured together by tapped aluminum angle pieces in the lower section.

Cut the aluminum angle pieces to length. Drill three ⁷⁄₃₂ in. diameter holes, and tap for ¼ in. threads. Bore corresponding ¼ in. holes through the plywood bottom of the upper section, and countersink in the top. Also drill three ¹¹⁄₆₄ in. holes in the opposite leg of the angle. Countersink. Fasten the two sections together with flathead machine screws. Fasten the aluminum angle pieces to the cherry wood sides with No. 8 by ½ in. flathead wood screws.

Finishing Touches

The inlaid medallion in the face of the upper section is a touch that gives just the right amount of decoration to the chest front. Craftsman Wood Service has a large variety of inlays to choose from, including classic decorations, fraternal emblems and zodiac signs. Incise the outline of the medallion. Then carefully rout the area to the thickness of the inlay, usually ¹⁄₂₈ in. deep. Glue the medallion in place, and sand smooth with 180 grit paper.

Inlaid initials are typical on tool chests to identify the owner. They also demonstrate the owner's woodworking skill. Rick inlaid his three initials in ⅛ in. thick ebony. Or con-

sider the monogram style, in which the last name initial is large and centered between the smaller first and middle initials. Rick used a sheet of Transartype T 1459 to design the letters. Large art supply stores have many examples of other type styles to choose from.

Apply letters to the case and to the ebony. Then carefully rout the waste wood in the case ⅛ in. deep, using an X-Acto knife and fine chisels. Cut out the ebony letters with a fine jigsaw blade, and clean the edges with appropriate size wood chisels. The inlay should be slightly raised when glued in place. Let it dry 24 hours and sand flush, using a hard-surface sanding block.

Stain as desired. Then finish the entire case, inside and out, with Watco natural finish oil. This finish goes on quickly and has a hand-rubbed appearance. Subsequent scratches or nicks can be easily repaired by sanding with 220 paper and re-oiling the area.

Apply oil liberally. Let it soak in overnight, then wipe off the excess. Apply a second and third coat as needed to achieve the desired richness of finish.

Line the inside of the drawers with felt, if desired. Although self-adhering felt is easily applied and protects fine tools, saw dust clings to it so it is difficult to clean.

Complete the case by installing an engraved nameplate on the saw holder. A plate can be ordered by mail from Creative Awards by

LANE, 32 West Randolph Street, Chicago, Illinois 60601 either with self-adhering back or with screw holes.

Secure a 28 in. level in front of the drawer-lock plunger piece with blocks at either end. Fit wood and metal retainers for other tools as desired.

Install furniture glides on the bottom of the case to nest into recesses in the top of the dolly.

Dolly

The dolly was made of oak with large, heavy-duty casters so it can also be used as a tool for moving casegoods.

The lap joint is simple, yet strong. The ends of the narrower boards extend 1 in. so that any checks won't readily develop into splits back to the carriage bolt holes.

Cut the pieces to size and clamp together. Mark the location of the holes for casters and bore them. Bolt the boards together with ¼ in. carriage bolts. Make two pieces of ¼ in. by 1 in. flat stock to lock the rear casters and prevent them from swiveling.

Use a piece of carbon paper to mark the location of the furniture glides on the bottom of the tool box. Cut recesses in the top of the dolly to the size of furniture glides. All the materials necessary for the chest are in the Craftsman Wood Service catalog, 1735 W. Cortland Ct., Addison, IL 60101.— *by David Warren. Photography by Matt Doherty.*

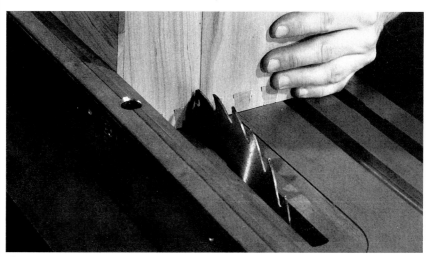

9 *Finish and assemble upper half of chest. Then saw the lid free, splitting the dovetails.*

Bath Organizer

Impress your guests with this eye-catching bath organizer.

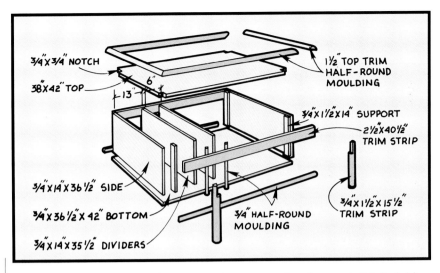

3/4"x3/4" NOTCH

38"x42" TOP

6"

13"

1½" TOP TRIM HALF-ROUND MOULDING

3/4"x1½"x14" SUPPORT

2½"x40½" TRIM STRIP

3/4"x14"x36½" SIDE

3/4"x36½"x42" BOTTOM

3/4"x14"x35½" DIVIDERS

3/4" HALF-ROUND MOULDING

3/4"x1½"x15½" TRIM STRIP

How many times have you wished you were better organized? With this bath organizer, you won't have to go dashing around for towels the next time guests show up. It's perfect for storing towels and bath accessories, and its clean lines make it as attractive as it is functional. Using just basic power tools, you can build the bath storage unit from ¾ in. plywood and half-round moulding.

Tips

A belt sander is very useful for finishing this project. Because the bath storage unit will be a showpiece for the bathroom, it is important that you buy pre-sanded plywood to reduce patching and additional hand sanding. Sink all nail heads and fill in nail holes with a nontoxic wood filler.

Construction

Begin by cutting the top, bottom and side pieces to their proper sizes. Use a straightedge on your circular saw to make the cuts. Equip the saw with a plywood cutting blade and cut with the plywood's good side down. Notch the four corners of the top piece with a saber saw.

Cut the supports and secure them to the side pieces with carpenter's glue and 3d finishing nails. Then secure the top and bottom to the side pieces with carpenter's glue and 6d finishing nails. Make sure to sink all nail heads.

Square the assembly and allow the glue to cure for 24 hours.

Cut the dividers to their proper size, and attach ¾ in. half-round moulding to the two edges with

carpenter's glue and 3d finishing nails. Position the dividers within the assembly and mark the areas to be cut away so the trim strips will fit properly. Make the slight notches with a saber saw. Check to insure that nails driven in the area to be cut are sunk adequately. Now secure the dividers in place with 6d finishing nails.

Glue and nail the 2½ in. wide trim strips in position. Secure them with 6d finishing nails. Cut the half-round trim strips to length on your power miter saw. Notch the top to fit around the adjacent trim strips. Secure these in place with glue and 6d finishing nails. Similarly, apply the top trim, which is half-round moulding. Cut the trim pieces to fit the top, and miter each edge to fit.

Complete the main project by attaching half-round moulding to the plywood bottom.

Finishing

Sink all nail heads and fill in all gaps with a nontoxic wood filler. Use a belt sander to smooth out the corners and flat surfaces. Then finish sand with a pad sander.

Carefully remove the dust and apply a sanding sealer. An airless spray gun is ideal for this type of application. When this coat has dried, sand lightly.

Apply your final finish and allow it to dry overnight. Then give it a light sanding and apply a second and final coat. — *by Cheryl Clark.*

This project is courtesy of Georgia-Pacific Corporation.

Lattice Gazebo

Add a touch of elegance to
your backyard with this traditional gazebo.

Anyone who owns a backyard yearns for the tranquility of an old-fashioned gazebo. This redwood gazebo features lattice panels to provide a sense of intimacy and a cedar shake roof so that you can use it on rainy days. Made from beautiful and long-lasting redwood, the gazebo requires a building area of only 12 ft. by 12 feet.

Tips

Before you undertake a project of this magnitude be sure you check with your building inspectors to obtain any necessary building permits. Modification of the construction plan may be required.

Use noncorrosive fasteners, such as galvanized nails, and make sure the concrete footings or piers are correctly positioned and leveled.

Construction

Start by building the base of the structure. Cut eight 2 x 8 redwood skirt components to the lengths shown in the diagram. Miter both ends of each board at a 22½ degree angle on your stationary saw.

Assemble the skirt components on a flat surface, and fasten with 16d galvanized finishing nails, three per joint. Assemble the 2 x 8 joists next. Start with one full-length joist across the width of the base. Cut the

ends of the joist to match the angles on the skirt.

Add two half sections of the joist material, perpendicular to the first joist, and nail the joists with 16d galvanized finishing nails. Finally, add the remaining four joist sections. Cut a double miter at both intersections of these joist members. Measure each joist individually to allow for variations in the base.

Before adding the decking, prepare a level site in your yard for the gazebo. Place concrete piers to be centered under each skirt joint and at the center of the base. Level these carefully, then set the base on the piers, securing it with lag screws

A combination of open spaces and lattice panels makes this redwood gazebo an intimate gathering place for family and friends.

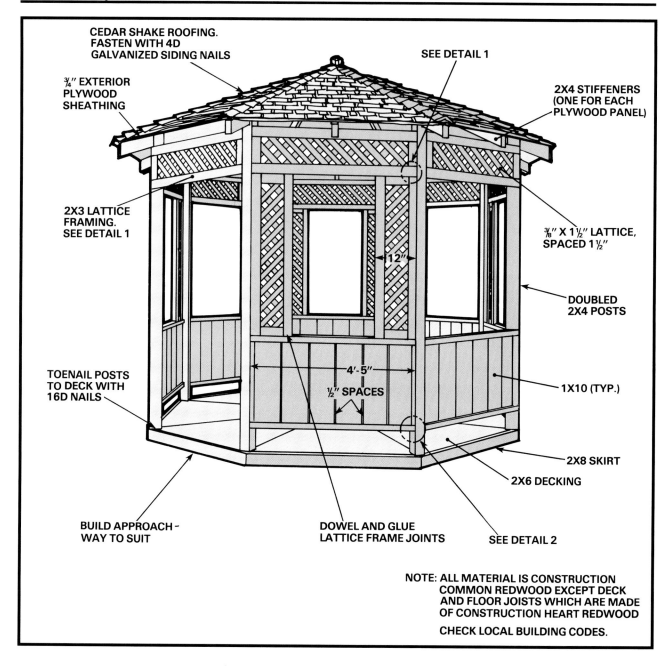

CEDAR SHAKE ROOFING.
FASTEN WITH 4D
GALVANIZED SIDING NAILS

SEE DETAIL 1

¾" EXTERIOR
PLYWOOD
SHEATHING

2X4 STIFFENERS
(ONE FOR EACH
PLYWOOD PANEL)

2X3 LATTICE
FRAMING.
SEE DETAIL 1

⅜" X 1½" LATTICE,
SPACED 1½"

12"

DOUBLED
2X4 POSTS

TOENAIL POSTS
TO DECK WITH
16D NAILS

4'-5"

½" SPACES

1X10 (TYP.)

2X8 SKIRT

2X6 DECKING

BUILD APPROACH –
WAY TO SUIT

DOWEL AND GLUE
LATTICE FRAME JOINTS

SEE DETAIL 2

NOTE: ALL MATERIAL IS CONSTRUCTION
COMMON REDWOOD EXCEPT DECK
AND FLOOR JOISTS WHICH ARE MADE
OF CONSTRUCTION HEART REDWOOD

CHECK LOCAL BUILDING CODES.

into lead anchors.

Now, set your table or radial arm saw to cut a 22½ degree miter, and begin laying the 2 x 6 decking. Start at the center of the floor and work outward. Lay out and cut one piece of decking for each octagonal row, then use this piece as a pattern for the others.

Space the decking boards ¼ in. apart and fasten them with 16d galvanized finishing nails. To prevent the wood from splitting, drill pilot holes for the nails in the decking boards.

Mill a ¾ in. by ¾ in. groove in one side of both 2 x 3 and 2 x 4 stock to form the panel frames. Use a dado blade on your stationary saw to form the grooves, stopping the grooves as shown in the diagram. Finish the stopped ends with a sharp wood chisel.

Cut the 67½ degree bevels and miters as indicated in the diagram (Details 1 and 2). Be sure to observe the safety precautions recommended for your saw.

Now, use a doweling guide to bore dowel pin holes in the lattice frame components. Assemble the individual panels with ⅜ in. by 2½ in. dowel pins and resorcinol glue, adding lattice screen or 1 x 10 lumber as required. Clamp the assemblies with pipe-style gluing clamps.

Next, make the perimeter posts from two 8 ft. long 2 x 4s. Use a dado blade in your saw to form the ¾ in. by 4 in. rabbets on each post half, as shown in the diagram.

Attach the panels to each post half with No. 10 by 2¼ in. flathead wood screws. Countersink the screw heads below the surface of the wood.

Position the post and panel assemblies on the base of the gazebo, and join the post halves with three

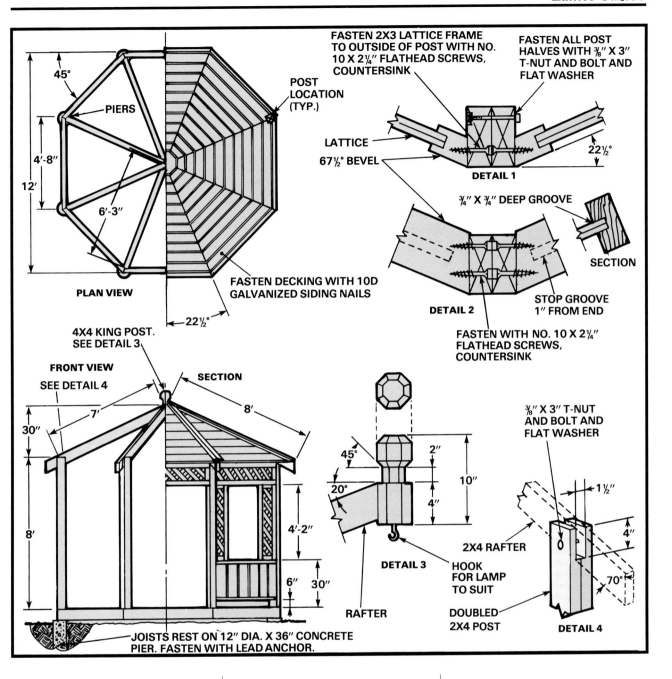

FASTEN 2X3 LATTICE FRAME TO OUTSIDE OF POST WITH NO. 10 X 2¼" FLATHEAD SCREWS, COUNTERSINK

FASTEN ALL POST HALVES WITH ⅜" X 3" T-NUT AND BOLT AND FLAT WASHER

POST LOCATION (TYP.)

LATTICE

67½° BEVEL

22½°

DETAIL 1

45°

PIERS

4'-8"

12'

6'-3"

PLAN VIEW

22½°

¾" X ¾" DEEP GROOVE

SECTION

STOP GROOVE 1" FROM END

DETAIL 2

FASTEN WITH NO. 10 X 2¼" FLATHEAD SCREWS, COUNTERSINK

FASTEN DECKING WITH 10D GALVANIZED SIDING NAILS

4X4 KING POST. SEE DETAIL 3

FRONT VIEW
SEE DETAIL 4

SECTION

7'

8'

30"

8'

4'-2"

6" 30"

⅜" X 3" T-NUT AND BOLT AND FLAT WASHER

45°

20°

2"

10"

4"

1½"

4"

DETAIL 3

2X4 RAFTER

HOOK FOR LAMP TO SUIT

70°

RAFTER

DOUBLED 2X4 POST

DETAIL 4

JOISTS REST ON 12" DIA. X 36" CONCRETE PIER. FASTEN WITH LEAD ANCHOR.

⅜ in. by 3 in. bolts and T-nuts. Move the segments slightly to equalize the octagonal shape. Use flat washers under the bolt heads. Toenail the posts to the base with 16d galvanized nails at the corners of the structure.

Now, make the king post (see Detail 3 of the diagram) from a length of 4 x 4 redwood. Start with a piece at least 2 ft. long. Set your saw to rip a 45 degree bevel, and rip off the corners of the post. Measure

This project is courtesy of the California Redwood Association.

carefully to provide equal widths on all sides of the octagonal profile. Cut the king post to length and form the finial on top with a rasp. Add the lamp hook, if desired.

Cut the rafters to the length and angles shown in the diagram. Nail two opposing rafters to the king post, then raise the assembly into position and bolt the rafters to the perimeter posts as shown in the diagram (Detail 4). Add the remaining rafters.

Measure and cut ¾ in. exterior plywood sheathing panels to fit the roof segments. Attach the central

stiffeners, then nail the sheathing to the rafters with 8d galvanized siding nails.

Finish construction by applying cedar shakes to the roof. Don't forget to add roof felt to serve as an additional moisture barrier.

Shingle each segment, then add a shake cap at each intersection. Take your time when shingling the roof to make a clean-looking pattern.

Protect the redwood structure with two or more coats of clear wood sealer and preservative. Allow 24 hours between coats. — *by George Campbell.*

Air Tool Cabinet

Keep all your air driven tools and accessories close at hand with this convenient tool cabinet/work center. The cabinet features full-extension drawers (above photo), removable overhead storage unit, retractable air hose reel, electric outlet strip and cork-covered top and drawers to protect tools. A red spray finish gives it the appearance of an automotive tool cabinet.

O ur tool cabinet was planned for those who rely on their air compressor and air tools to do home, shop and automotive do-it-yourself jobs. The basic cabinet shell, as well as many of the ideas used, are also easily adaptable to other shop needs.

Your air driven tools depend upon a central air source and should be organized efficiently to enable you to quickly release one tool and hook up another. This cabinet organizes all your tools and acces- sories and keeps the supply line at your fingertips. Once the air compressor is hooked up to the air hose reel, the cabinet can be moved the radius of the air compressor's supply hose (about 15 ft. or more). With the reel fully extended, you will be able

Store up to 500 lbs. of air driven tools and accessories in this portable cabinet.

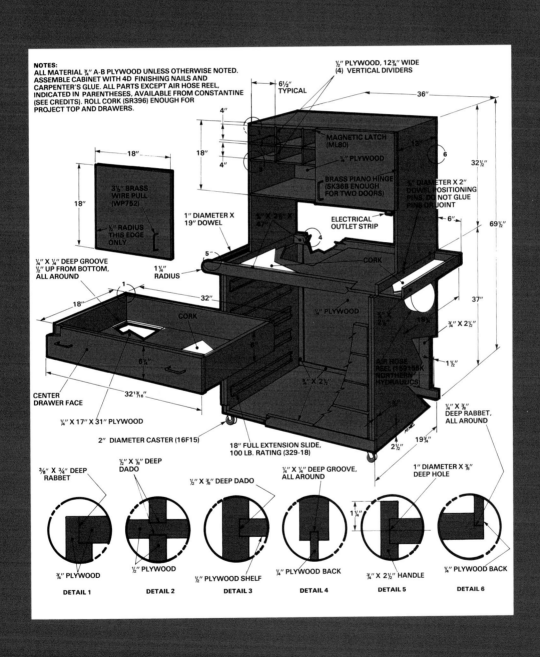

NOTES:
ALL MATERIAL ¾" A-B PLYWOOD UNLESS OTHERWISE NOTED. ASSEMBLE CABINET WITH 4D FINISHING NAILS AND CARPENTER'S GLUE. ALL PARTS EXCEPT AIR HOSE REEL, INDICATED IN PARENTHESES, AVAILABLE FROM CONSTANTINE (SEE CREDITS). ROLL CORK (SR396) ENOUGH FOR PROJECT TOP AND DRAWERS.

½" PLYWOOD, 12¾" WIDE
(4) VERTICAL DIVIDERS

6½" TYPICAL

36"

4"

18"

MAGNETIC LATCH (ML80)

¼" PLYWOOD

32½"

3½" BRASS WIRE PULL (WP752)

BRASS PIANO HINGE (SK36B ENOUGH FOR TWO DOORS)

¼" DIAMETER X 2" DOWEL POSITIONING PINS. DO NOT GLUE PINS OR JOINT

18"

1" DIAMETER X 19" DOWEL

ELECTRICAL OUTLET STRIP

6"

69½"

½" RADIUS THIS EDGE ONLY

5

CORK

¼" X ¼" DEEP GROOVE ½" UP FROM BOTTOM, ALL AROUND

1¼" RADIUS

1

37"

18"

32"

¼" PLYWOOD

¾" X 2½"

CORK

CENTER DRAWER FACE

6¼"

AIR HOSE REEL (159155K NORTHERN HYDRAULICS)

1½"

32¹³⁄₁₆"

¼" X 2½"

19¾"

¼" X 17" X 31" PLYWOOD

2" DIAMETER CASTER (16F15)

18" FULL EXTENSION SLIDE, 100 LB. RATING (329-18)

2½"

¼" X ¾" DEEP RABBET, ALL AROUND

⅜" X ¾" DEEP RABBET

½" X ⅞" DEEP DADO

½" X ⅞" DEEP DADO

¼" X ¼" DEEP GROOVE, ALL AROUND

1" DIAMETER X ⅞" DEEP HOLE

1¼"

¾" PLYWOOD

½" PLYWOOD

½" PLYWOOD SHELF

¼" PLYWOOD BACK

¾" X 2½" HANDLE

¼" PLYWOOD BACK

DETAIL 1 **DETAIL 2** **DETAIL 3** **DETAIL 4** **DETAIL 5** **DETAIL 6**

Construction Details

Make the grooves in the 1 x 3s to accept the back (base unit). Using your table saw, make two passes with a 1/8 in. blade, or use a 1/4 in. dado head.

Cut rabbets in the drawer sides on a table saw using a wobble-type dado blade.

Notch all four corners of the double-layered sides (lower unit) with a saber saw. Cut with the good side up on the table saw to avoid splintering the good surface.

Attach the doubled-layered sides to the top, upper front rail and double-layered bottom. Make sure you square the cabinet. Then cut and add the lower rail.

Assemble the back rails, stiles and the handle, then attach the casters.

To assemble the drawers, glue and nail a side to the front and back, then slide the drawer bottom into the grooves. Glue and nail the remaining side.

Assemble the top unit with glue and 4d finishing nails. An air driven brad nailer speeds assembly. If assembling with clamps, follow the steps outlined in the text. Wear eye goggles when operating an air nailer.

to remove lugs from automobile tires about 43 ft. away from the air source.

Double-layered cabinet sides and bottom and 100-lb.-test drawer glides allow the cabinet to store up to 500 lbs. of heavy tools and accessories.

You can cut construction costs by substituting particle board for the 3/4 in. A-B plywood. We used plywood to make the project lightweight while providing durability.

Features

This cabinet has a long list of special features:

- Full-extension drawers mounted on 100-lb.-test glides hold 20 or more air tool accessories, with easy access to all.

- The drawers and work surface are lined with cork to protect your tools.

- All tools can be stored in drawers or behind doors to keep them free of dust, a plus in any workshop.

- The top storage unit holds smaller tools, hose connectors, lubricating oil and other supplies.

- The top unit is assembled to the base on dowel pins so it can be removed when you need more work space for large projects.

- Heavy-duty casters and a convenient handle let you move the cabinet to where it is needed.

- An air hose reel is attached to the cabinet and connected to your air compressor so you are always ready to work.

Lower Cabinet Construction

Cut all the project's parts to width. Next, cut to length everything but the drawer faces, drawer fronts and backs, 1 x 3 stiles and lower rails, the upper unit's dividers, shelves and top. Cut these to length only after you have cut the dadoes and rabbets and are assembling the unit, so everything fits properly.

Always cut plywood with a fine-toothed plywood blade, and keep the plywood's good side up on the table saw (to avoid splintering the good side.) Use plenty of carpenter's glue and 4d finishing nails or air driven brads during assembly.

Make up the double-layered sides and bottom by gluing and securely clamping them together. Pay careful attention to how these layers interconnect adjacent pieces. For instance, the side's inner layer is ¾ in. shorter on the bottom.

Next, cut out the four corner notches in each side with a saber saw. Again, use a plywood cutting blade.

Draw the radius on the ends of the upper front 1 x 3 rail handle with a compass. Then cut out the radius with a saber saw. Next, drill counterbored 1 in. diameter holes into the handle ends with a drill equipped with a spade bit.

Attach the two sides to the bottom and (lower cabinet) top. Assemble the unit with the front side up. Square the unit and attach the upper front 1 x 3 rail. Cut and install

the lower rail.

Now turn the unit so the back side is facing up. Lay the upper and lower back rails in place and mark them for grooving. Cut stopped grooves into the rails and full grooves into the stiles with a table saw.

Affix the back stiles and rails along with the plywood back and dowel handle pull. Make sure this side of the cabinet is also square. Now install the casters and place the unit upright.

Rabbet all the drawer sides on your table saw using a wobble-type dado blade. Next, lay out the drawer sides and mark the fronts and backs to fit.

Cut the fronts and backs to length and groove all the drawer sides to accommodate the plywood bottoms of the drawers. Assemble the drawer sides along with the bottoms. Square as you go.

Install the drawers along with the full-extension slides. Measure and cut the drawer faces to fit the exact opening.

Mark the positions for the drawer pulls on the drawer faces and drill the holes for the mounting screws. Counterbore the inside screw locations.

Upper Unit Construction

Dado and rabbet the upper unit's sides, then cut its top, bottom and shelves to fit. Cut dadoes and rabbets into the top and the bottom shelf. Cut stopped rabbets into the two sides to hold the plywood back.

Make a dry assembly and mark the divider locations. Remove the shelves and cut dadoes for the dividers on a table saw.

Cut the dividers to fit, and assemble the upper cabinet with glue and nails. Square the assembly before attaching the ¼ in. plywood back.

Set the upper unit on the cabinet and mark two dowel locations. Drill holes for these positioning dowels.

Rout a ½ in. radius on the outside edge of both doors and install the doors to the unit with piano hinges.

This project is courtesy of Workbench magazine.

Also locate and drill holes for the brass wire drawer pulls. Now remove the piano hinges for final finishing.

Finishing Touches

Make sure all nails are recessed. Fill in any blemishes with wood filler and finish sand the entire project.

Before painting, remove all drawer and door hardware.

Thoroughly wipe off all dust and then spray paint the cabinet with a latex enamel primer. When dry, lightly sand this coat and apply a final layer of red latex enamel.

Install the brass wire drawer pulls, piano hinges, magnetic latches, drawer fronts, outlet strip and air hose reel. Cut and lay the cork onto the lower cabinet's work surface and drawers.

Finally, insert the positioning dowel pins into the lower cabinet and set the upper cabinet in place. — *by Al Gutierrez.*

Sources:

Air tools: Ingersoll-Rand, P.O. Box 241154, Charlotte, NC 28224.

Air tools: Stanley Bostitch, 420 South Kitley Ave., P.O. Box 19329, Indianapolis, IN 46219.

Air tools: Sanborn Mfg., 118 West Rock St., Springfield, MN 56087.

Air tools: Campbell Hausfeld, 100 Production Dr., Harrison, OH 45030.

Air tools: Sears catalog.

Air tools: Mark 1, distributed by Trade Associates, P.O. Box 1522, Kent, WA 98032.

Brass wire pulls, drawer glides, hinges, latches, cork, casters: Constantine, 2050 Eastchester Rd., Bronx, NY 10461, 1-800-223-8087 (1-800-822-1202 in New York).

Retractable air hose reel: Northern Handyman, Inc., P.O. Box 1499, Burnsville, MN 55337, 1-800-533-5545 (612-894-8310 in Minnesota).

Modular Table and Chairs

Why settle for an ordinary kitchen table and chairs when it is so easy to create your own to perfectly match your kitchen's decor?

This project is courtesy of Georgia-Pacific Corporation.

The kitchen was once the center of family life. The warmth of the hearth created the right atmosphere for gathering for meals or just plain visiting with family and friends. But with today's busy schedules, there is little time for family meals. Yet, the kitchen is the highest traffic area in most homes. Family members come and go, grabbing a snack, fixing a bite of lunch or making something quick for dinner. This modular table and chairs is designed for today's busy family, yet preserves the idea of the kitchen as a natural gathering place.

Tips

Use carpenter's glue and finishing nails to assemble the project. Pre-drill holes for the nails located at the edges of the mitered plywood.

Buy only quality plywood that is warp-free and sanded on both sides.

Construction

Carefully measure and cut out the top with a circular saw equipped with a plywood blade. Guide the circular saw along the cutting line with a straightedge clamped to the work surface. Cut with the good surface placed on the bottom. This prevents excessive wood tear.

Next, rabbet the edge strips from long lengths of material on your table saw. Then smooth out any rough spots in the rabbet cut with a sharp wood chisel. Attach the edge strips to the table top with glue and 4d finishing nails, countersinking all nail heads.

Now cut the crosspieces and braces to their proper lengths. Glue and secure the crosspieces to the bottom of the table top with No. 8 by 1¼ in. flathead wood screws (countersunk). Double check the measurements for the plywood panels and end strips. Then cut each of these pieces to size and miter the joining edges carefully. Use a circular saw set to a 45 degree miter and guide the saw along the straightedge. Check the quality of the miter by cutting some scrap material first.

Assemble the base plywood panels to the crosspieces with glue and 4d finishing nails. Remember to also glue and nail the plywood panels and end strips to one another. Finish the base unit by gluing and nailing the braces within the bottom of the assembly. Again use 4d finishing nails.

Now cut and miter the side panels for each of the four seats. Then cut the corner blocks and secure them to the side panels with glue and 4d nails driven into the side panels.

Custom cut the seats and install with glue and 4d finishing nails.

If you are going to spray paint the project, it is best to do so at this time, before applying the top's laminate.

Lamination

Laminate the top and top edges by first applying the edges. To do so, oversize the laminate by ¼ in. in width and length, and cut it out using a saber saw equipped with a metal or plastic cutting blade. Use a coarse blade to prevent chipping the laminate.

Next, apply a contact adhesive to two opposite edges of the table top and affix the laminate. Use a J-roller to smooth the laminate to the table's surface. Then use a flushing bit installed in your router to trim each of these two edge strips.

Similarly, apply contact ad-

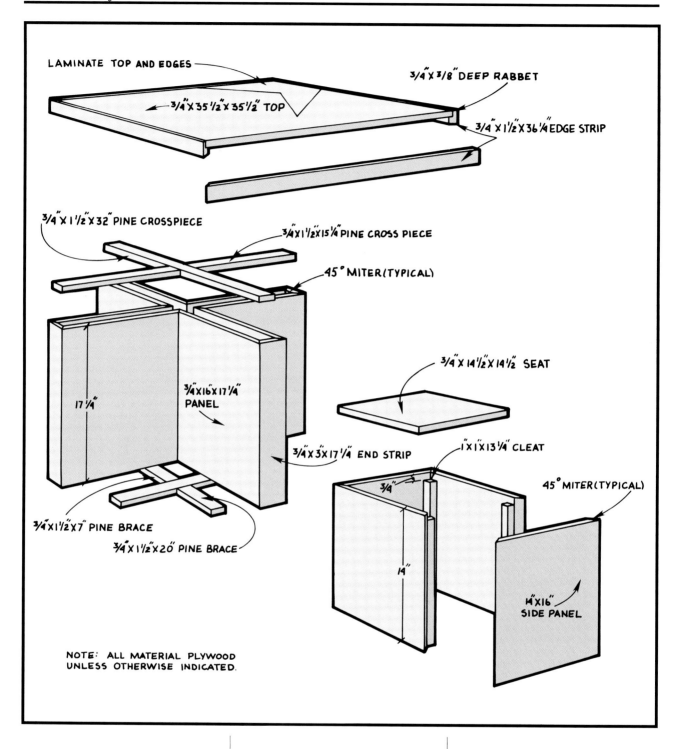

LAMINATE TOP AND EDGES

¾"X 35½"X 35½" TOP

¾"X ⅜" DEEP RABBET

¾"X 1½"X 36¼" EDGE STRIP

¾"X 1½"X 32" PINE CROSSPIECE

¾"X1½"X15¼" PINE CROSS PIECE

45° MITER (TYPICAL)

17¼"

¾"X16"X17¼" PANEL

¾"X 14½"X 14½" SEAT

¾"X3"X17¼" END STRIP

1"X1"X13¼" CLEAT

45° MITER (TYPICAL)

¾"

14"

¼"X16" SIDE PANEL

¾"X1½"X7" PINE BRACE

¾"X 1½"X 20" PINE BRACE

NOTE: ALL MATERIAL PLYWOOD UNLESS OTHERWISE INDICATED.

hesive to the laminate and to the two remaining edges. Smooth the edges with the J-roller and trim the excess with your flushing bit.

Cut the laminate for the top ¼ in. oversize, and apply adhesive to both the wood surface and the laminate. If there are any impurities in the surface of the table top, use a good wood filler before applying the adhesive. After installing the laminate, use a J-roller to insure good adhesive contact. Likewise, use a trimming bit to dress the top. If you do not have a trimming bit, use a file and carefully bevel the top edges. Do not file into the laminate's edge, but away from it.

Carefully sand all of the remaining parts with a pad sander, and break all edges with a sanding block. Wipe away the loose dust with a damp cloth.

Apply a sanding sealer and lightly sand after it has dried. Apply two coats of paint in the color of your choice, sanding lightly between coats.

Install nylon glides on the bottoms of the chairs and the table to prevent damaging your kitchen floor or the edges of the plywood. — *by Jonathan Wesley.*

Tile Top Country Coffee Table

This elegant-looking coffee table is a simple project. It uses ready-made legs and ceramic tile.

Building this great-looking table is a lot less complicated than it appears. The ready-made legs save time and effort. Applying a light tan ceramic top to the "scrubbed look" trestle makes it attractive and durable.

The size of the tiles selected determines the size of the top. This project uses nine No. AK-21 sheeted tiles from Color Tile that measure 12¼ in. square. Each sheet contains four 6 in. tiles separated by a ¼ in. grout space. Select an alternate color if desired.

Cabriole Legs

The legs measure 14⅝ in., the appropriate height for a coffee table. Cut ⅝ in. deep mortises from two directions in each leg by hand or with a mortising jig.

Side Frame

Next, cut four side pieces from 1 x 6 knotty pine. Cut each to length. Use a hand saw or tenoning jig to cut ⅝ in. tenons to fit the leg mortises. Mark mating joints.

Then cut the scallop design on each of the four side pieces. Sand curves smooth.

Miter the four 2 x 4 supports for the plywood. Glue and nail to the pine side pieces so that the 2 x 4s are 1½ in. below the top edges of the sides.

Assemble the legs and sides with white glue, and clamp until the glue has dried. Be sure the assembly is square; check by measuring diagonally in each direction. When both dimensions are identical, the unit is square.

Plywood Base

After the glue has thoroughly dried, cut two pieces of ¾ in. plywood to fit the inside dimension of the table.

Apply white glue to the 2 x 4 frame and set in one sheet of plywood. Use No. 10 by 1¼ in. flathead screws to secure the sheet to the frame.

Then apply glue to the surface of the plywood, and lay in the second sheet. Run several screws of the same size through the top sheet and into the lower one to lock them tightly together. This insures a flex-free surface on which to apply ceramic tile. Also, around the border, run several No. 10 by 2¼ in. flathead screws through the plywood pieces and into the 2 x 4 supports. It is important to countersink all screws.

Ceramic Tile Top

Allow the glue to set, then spread on a ceramic tile mastic. If desired, draw a grid pattern on the plywood to guide the tile laying. When

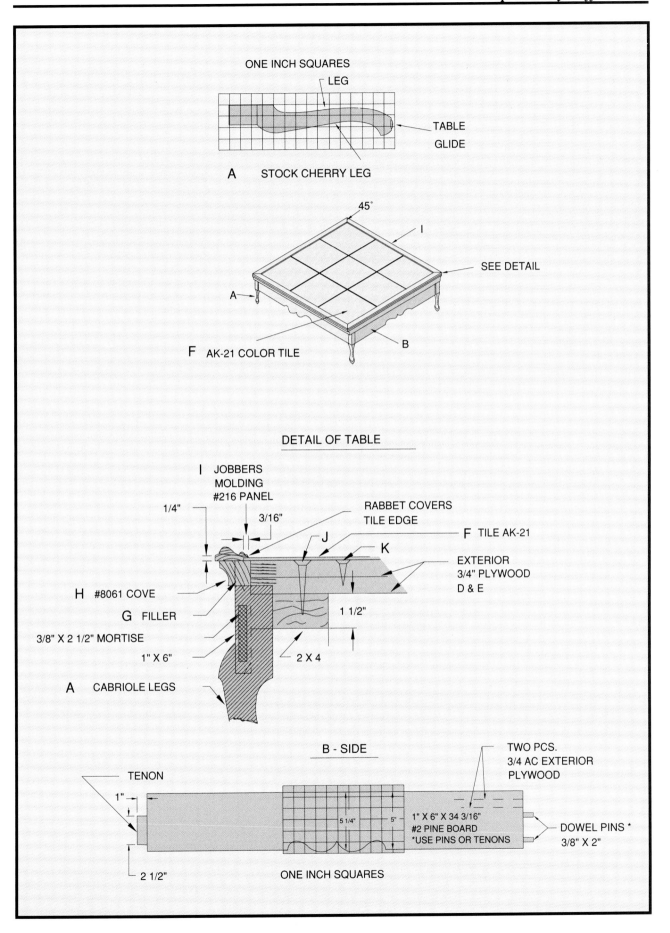

ONE INCH SQUARES

LEG

TABLE

GLIDE

A   STOCK CHERRY LEG

45°

I

SEE DETAIL

A

B

F   AK-21 COLOR TILE

DETAIL OF TABLE

I   JOBBERS
MOLDING
#216 PANEL

1/4"

3/16"

RABBET COVERS
TILE EDGE

J

K

F   TILE AK-21

EXTERIOR
3/4" PLYWOOD
D & E

H   #8061 COVE

G   FILLER

1 1/2"

3/8" X 2 1/2" MORTISE

1" X 6"

2 X 4

A   CABRIOLE LEGS

B - SIDE

TWO PCS.
3/4 AC EXTERIOR
PLYWOOD

TENON

1"

5 1/4"   5"

1" X 6" X 34 3/16"
#2 PINE BOARD
*USE PINS OR TENONS

DOWEL PINS *
3/8" X 2"

2 1/2"

ONE INCH SQUARES

113

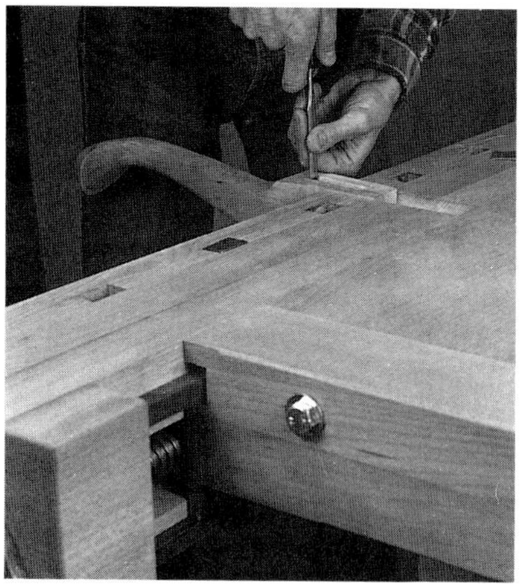

***1** Richly-carved cherry Cabriole legs require cutting mortises only for tenoned sides before assembly. Cut by hand or with a mortising jig on a drill press.*

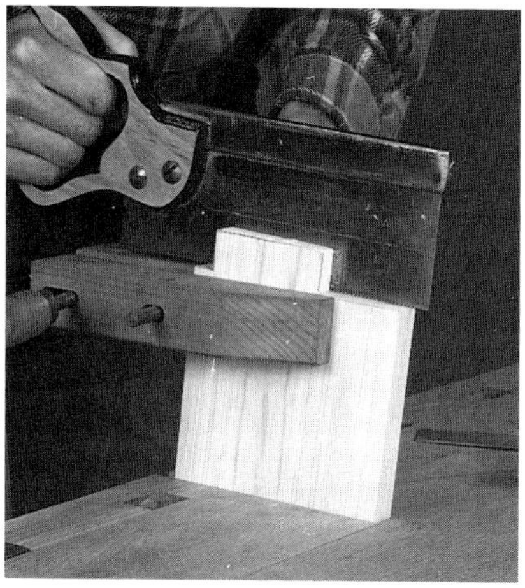

***2** Tenons can also be cut by hand, or with a jig on a table saw. A hand screw clamp is a useful guide for the saw when making cross-grain cuts.*

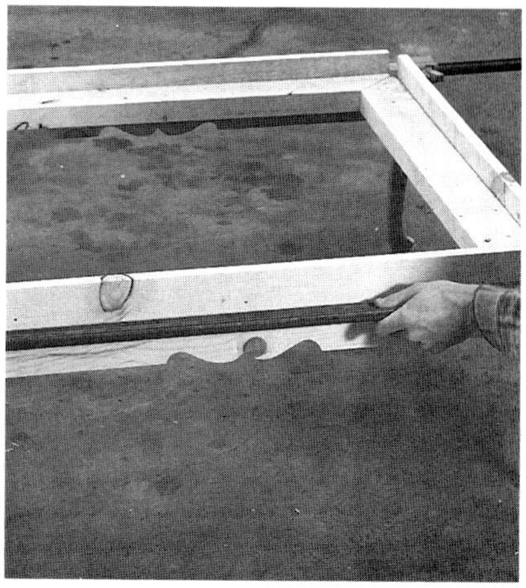

***3** Glue and nail 2 x 4s to the side frame pieces. Then coat with glue and fit the tenons into the mortised legs. Clamp the assembly until the glue has dried.*

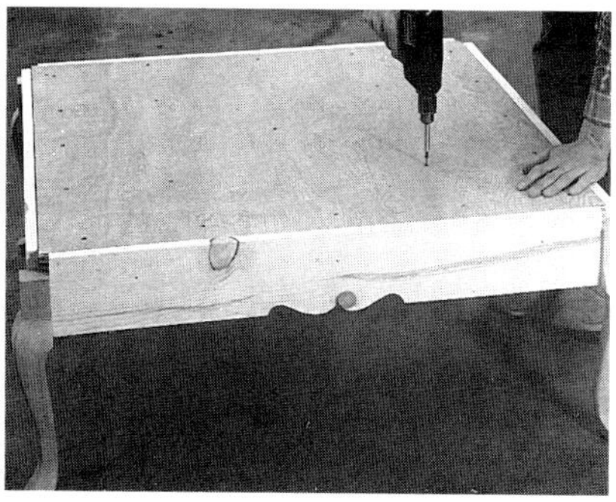

***4** Glue and screw the lower sheet of plywood to the 2 x 4s with No. 10 by 1 1/4 in. flathead screws. Then, using No. 10 by 2 1/4 in. flathead wood screws, fasten the top layer to the bottom sheet and to the 2 x 4s.*

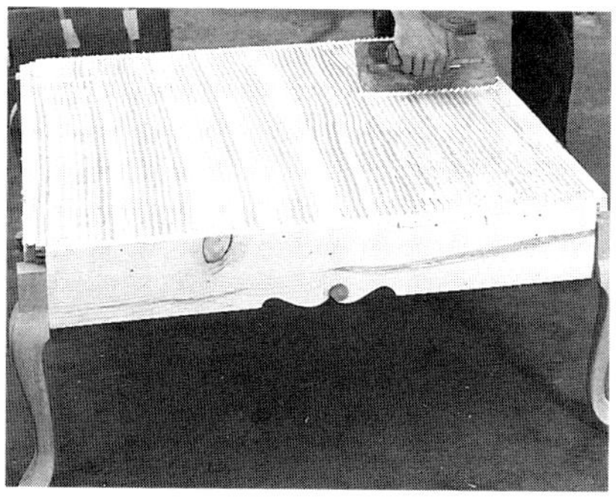

***5** Use a 3/8 in. V-groove trowel to spread mastic over plywood. If alignment lines are drawn, leave these exposed so they can be followed.*

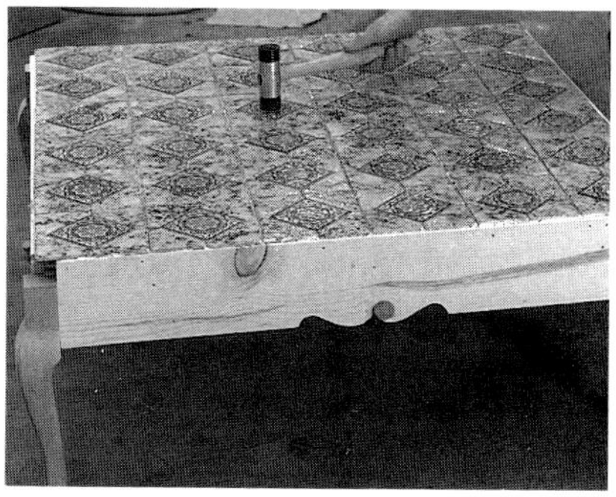

***6** Place tiles onto the sticky mastic surface, making certain the edges are straight and that the spaces between tiles are equal. Tap lightly with a rubber mallet to assure proper contact with the mastic.*

spreading mastic, leave the lines exposed.

Lay the tiles. After the mastic has set, apply grout, filling the spaces between tiles completely. Allow the grout to dry. Then clean the surface, making sure that all traces of grout are removed. Carefully brush a silicone sealer onto the grout lines to protect them from dirt.

Top Moulding

Next, apply filler strips to the top edge of the side pieces. Then nail and glue cove moulding over the filler strips. Cut a 1/4 in. by 3/16 in. rabbet on the flat side of the panel moulding per the drawing, and apply strips measuring 9/16 in. by 1 1/2 in. by 39 1/2 in. on top of the cove, filler and edge of the side pieces.

BILL OF MATERIALS — Tile Top Country Coffee Table

Finished Dimensions in Inches

A	Leg	1 11/16 x 1 11/16 x 14 5/8 cherry	4
B	Side	3/4 x 5 x 36 3/8 pine	4
C	Base Support	1 1/2 x 3 1/2 x 36 1/4 fir	4
D	Upper Base	3/4 x 3/4 x 36 1/4 exterior plywood	1
E	Lower Base	3/4 x 3/4 x 36 1/4 exterior plywood	1
F	Table Surface	1/4 x 6 x 6 ceramic tile	36
G	Filler Strip	3/8 x 1 x 38 1/2 pine	4
H	Side Edge	11/16 x 1 1/2 x 39 7/8 pine cove	4
I	Top Edge	9/16 x 1 1/2 x 39 7/8 pine panel moulding	4
J	Fastener	No. 10 x 2 1/4 flathead screws	16
K	Fastener	No. 10 x 1 1/4 flathead screws	5

7 *When the mastic is dry, fill the spaces between the tiles with grout using a spatula. Grout dries in about 20 minutes, so mix only enough to work with at a comfortable speed. One-third of this table was grouted at a time. Wipe away excess grout; when thoroughly dry, clean with a coarse, dry cloth.*

Finish

Resand the assembly with 120 grit paper, particularly where grout moisture may have gotten onto the pine frame.

To finish, match the cherry legs to the pine. In this project, two coats of Minwax platinum stain were applied to the legs to bring the color close to the pine finish. Next, coat the entire assembly with a stain, such as Minwax cherry oil stain used in this project. When the stain has dried, a coat of dark walnut oil stain was applied. Finally, Minwax platinum stain was applied to the grooves and cracks to give the table the appearance of a refinished antique. To make the table look even more like an antique, use a chain to distress the surface of the wood with tiny dings and dents.

Fill nail holes with a Weldwood wax stick of an appropriate color. Then apply two coats of satin varnish or polyurethane.

When the finish has dried, wipe on a coat of carnuba wax to further protect your handsome new table.

Full-size plans are available for $9.95 from Creative Features, Coffee Table, Suite 1317, Six North Michigan Ave., Chicago, IL 60602. — *by William Beyer. Photography by William Laskey and David J. Warren. Design by Virginia Howley, I.D.S.*

8 *Use a damp cloth to smooth grout joints to uniform levels for an attractive appearance. After this step, wipe away as much excess grout from the surface of the tile as possible.*

9 *Wipe grout film from the tile surface with a damp rag. Rinse the rag frequently in clear water and continue wiping until the tile sparkles.*

10 *Finally, miter rabbeted panel moulding; apply it over the filler strip and the top edge of the side pieces.*

Butcherblock Bench

Bring style to your outdoor living space with this redwood bench.

This redwood butcherblock bench is a stylish addition to decks or garden patios. The laminated redwood bench is 18 in. high by 16½ in. wide by 48 in. long. If desired, you can build yours to a different length.

Tips

Because the project will need to withstand heat, cold and water, it is important to use a suitable weather-resistant wood. Construction Heart should be used for the legs (where increased decay resistance is needed) and Construction Common

and Merchantable may be used for seats.

Use noncorrosive nails and hardware such as stainless steel, aluminum or hot-dipped galvanized.

Construct the project on a flat surface to insure that the legs will rest squarely on the surface when the bench is moved to the patio or deck. To prevent splitting the wood, predrill nail holes near the ends of the boards.

Construction

The entire project is comprised of various lengths of 2 x 4s. Cut all of

the pieces to length on a radial arm saw or power miter saw.

To facilitate construction, assemble short and long pieces in pairs and secure them with 8d nails. Drive the nails in at an angle to help secure the workpieces.

Secure one section of legs to a section of the top and use a carpenter's square to insure that the assembly is square. Continue adding a pair of legs to a section of top until you have completed the layers. With each layer, use a carpenter's square to make sure that the bench top is absolutely flat.

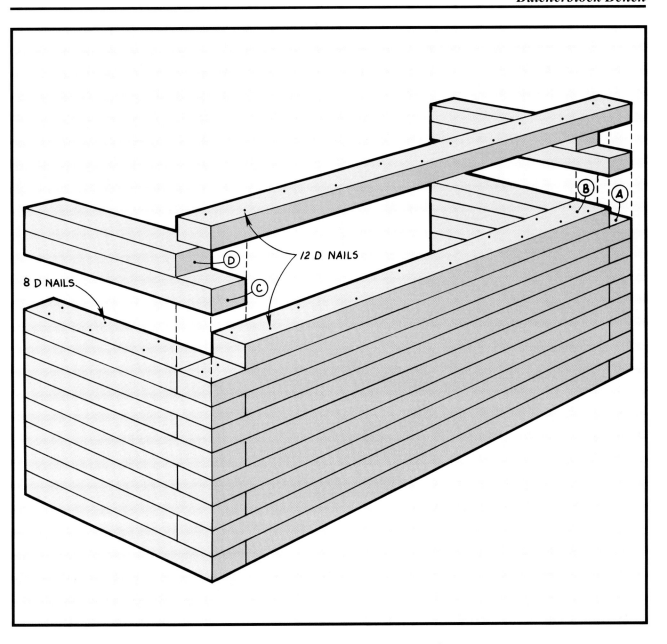

8 D NAILS

12 D NAILS

Ⓓ

Ⓒ

Ⓑ Ⓐ

Note that this project does not require that you joint the edges of the 2 x 4s before constructing the materials. However, to obtain a nice solid-looking bench, use a stationary jointer or portable hand planer to dress the wood's edges before assembling the workpieces.

Finishing

Sand the project with a belt sander and finish the hard-to-reach areas with a hand sander. Then dull all the sharp edges with a sander or hand plane. Because redwood is so soft,

This project is courtesy of the California Redwood Association.

you need to be careful not to sand into the wood's grain. Otherwise, the wood may split.

You can apply a finish or leave the wood to weather naturally. Water repellents are recommended for a natural appearance and as an

undercoating for some finishes. Refrain from using toxic materials or chemicals. Pigmented stains can be used to provide additional color to the warmth of the redwood. — *by Jonathan Wesley. Photo by George Lyons.*

BILL OF MATERIALS — Butcherblock Bench

Finished Dimensions in Inches

A	Top	$1\frac{1}{2}$ x $3\frac{1}{2}$ x 48 redwood	5
B	Top	$1\frac{1}{2}$ x $3\frac{1}{2}$ x 41 redwood	5
C	Long Leg	$1\frac{1}{2}$ x $3\frac{1}{2}$ x 18 redwood	10
D	Short Leg	$1\frac{1}{2}$ x $3\frac{1}{2}$ x $14\frac{1}{2}$ redwood	10

Child's Table

This delightful small table has a pine top with cherry wood legs and stretcher.

The original table is all pine, but cherry wood makes a stronger trestle. A handsome combination, the table can have many uses in today's home. However, in its mid-Victorian period, this was a child's work table for school homework. Both top and stretcher took abuse from pencils, spilled ink and squirming feet. So a pine top was a logical choice because it was inexpensively replaced when badly marred. Convenient screws from the top hold both members.

To reproduce the table, make a 1/8 in. Masonite template of the leg pattern and lay it out on cherry wood. Cut about an inch longer than required. Do the layout on two boards. Joint the edges to be glued using a try plane or jointer, and bore 1/4 in. diameter holes for dowel pins per the drawing. Join the four boards in pairs by screwing them together in areas outside the pattern. Cut them out on a band saw, or use a classic chairmaker's saw. Sand the edges with a band sander, or by hand, using 80 grit paper followed by 120 grit sandpaper. Use a 1 1/4 in. drum sander or scraper for tight spots. After sand-

ing, separate boards and cut away the excess stock. Save all cutoffs for use when clamping glued parts. Then apply glue to the dowel pins, insert them into the bored holes, glue the edges, and join and clamp the leg parts.

Next, lay out the patterns for the pine top and cherry stretcher. Allow 1/16 in. outside the width lines when cutting and sand to the line for accuracy. With a try plane, joint the edges of the top to be glued, and bore dowel pin holes per the drawing. To restrain warping, alternate the grains

in the two pieces, then glue and clamp. When dry, round the edges of the top and stretcher with a 1/4 in. router bit or spokeshave. For small battens, use scrap from the stretcher before thicknessing the wood to 3/8 inch. Saw large battens from 1 1/4 in. square cherry stock per the drawing and sand smooth. Sand all parts with 80 grit paper and then with 120 grit sandpaper.

Assembly

Use a 3/8 in. plug cutter to cut pine and cherry plugs from cutoffs.

BILL OF MATERIALS — Child's Table

Finished Dimensions in Inches

A	Top	1/2 x 7 3/8 x 23 9/16 pine	1
B	Top	1/2 x 8 1/8 x 23 9/16 pine	1
C	Stretcher	3/8 x 8 x 19 1/2 cherry	1
D	Leg	5/8 x 6 1/8 x 22 3/4 cherry	4
E	Upper Batten	7/8 x 1 x 12 3/4 cherry	2
F	Lower Batten	1/2 x 3/4 x 8 cherry	2

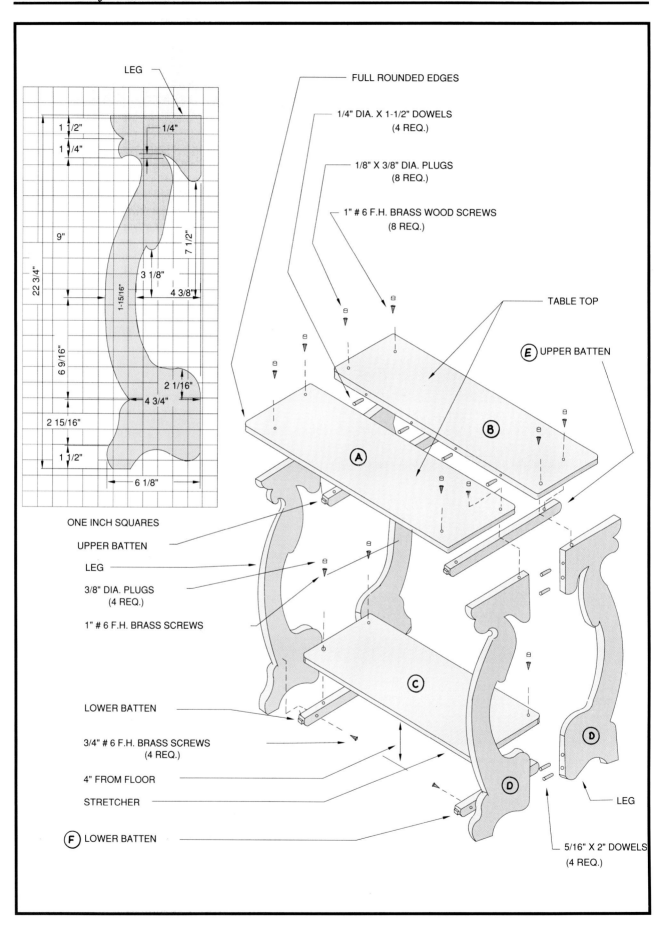

LEG

FULL ROUNDED EDGES

1/4" DIA. X 1-1/2" DOWELS
(4 REQ.)

1/8" X 3/8" DIA. PLUGS
(8 REQ.)

1" # 6 F.H. BRASS WOOD SCREWS
(8 REQ.)

TABLE TOP

Ⓔ UPPER BATTEN

1 1/2"
1/4"
1 1/4"
9"
7 1/2"
3 1/8"
1-15/16"
4 3/8"
22 3/4"
6 9/16"
2 1/16"
4 3/4"
2 15/16"
1 1/2"
6 1/8"

Ⓑ

Ⓐ

ONE INCH SQUARES

UPPER BATTEN

LEG

3/8" DIA. PLUGS
(4 REQ.)

1" # 6 F.H. BRASS SCREWS

Ⓒ

LOWER BATTEN

3/4" # 6 F.H. BRASS SCREWS
(4 REQ.)

4" FROM FLOOR

Ⓓ

STRETCHER

Ⓕ LOWER BATTEN

Ⓓ

LEG

5/16" X 2" DOWELS
(4 REQ.)

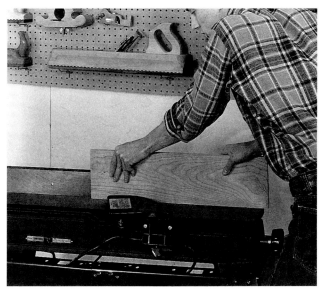

1 *Use either a jointer or a traditional try plane to dress edges for joining.*

2 *A hand try plane may be used to thickness boards for the table top, legs, stretcher and battens. Follow with a smoothing plane.*

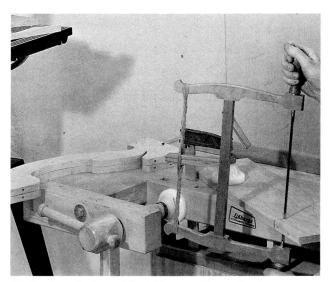

3 *To shape the legs of the table, use either a classic chairmaker's saw or a band saw. A ⅛ in. wide band saw blade is required for sharp curves.*

4 *With legs shaped, replace the band saw blade with a sanding belt to quickly smooth edges.*

Carefully mark and bore ⅜ in. diameter plug holes in the top, stretcher and small battens per the drawing. Bore plug holes ¼ in. deep in the top, ⅛ in. deep in the stretcher and ½ in. deep in the battens.

Then bore 3/32 in. diameter pilot holes through the stretcher and into the battens. Screw and glue the battens to the stretcher. Plug the holes with cherry plugs, matching grain direction. When the glue is dry, use a chisel with the bevel side down to trim off excess. Then sand the entire stretcher smooth.

Mark a line on the inside of the legs for the stretcher height. Mark 3/32 in. diameter pilot holes on the legs. Bore holes in the legs and battens. Fasten the legs to the stretcher with screws, then plug the holes.

Bore pilot holes through the top into the large battens. Glue and screw together per the drawing. Then place the top on the legs and screw it to the legs. Plug the holes and allow the glue to dry. Pare the plugs down, and sand the top and battens smooth.

Finishing

Finish sand the entire table and apply Minwax cherry stain to the top only. Let the stain dry overnight. Then apply three to five coats (allow to dry between coats) of Watco natural oil to the entire table. Cherry will mellow with age. Full size plans are available for $9.95 from Creative Features, Child's Table, Suite 1317, Six North Michigan Ave., Chicago, IL 60602. — *by David Warren.*

Freestanding Chopping Block

This sturdy chopping block is just what the cook ordered.

W hether it's for whipping up a quick meal or preparing a gourmet feast, there isn't a cook around who wouldn't appreciate a handy chopping block. This project features a solid top, measuring 3½ in. by 24 in. by 24 in. and a rugged frame that won't wobble. The top is made of maple, but birch can be substituted.

Tips

Select kiln-dried lumber that is knot-free. Do not in any case use a toxic adhesive since food comes in contact with the surface. Carpenter's glue is an acceptable adhesive for constructing the top.

Construction

Carefully crosscut the 16 workpieces making up the top. All of the workpieces must be absolutely square. If the surfaces to be contacted are slightly warped, you must flatten these on your jointer or thickness planer. If the mating surfaces are not flat, you will have a nightmare when it comes to assembling the top and there will be unsightly gaps. Cut a piece of scrap material on your saw and check the edge with a combination square before cutting the actual workpieces.

Now cut the four legs to their proper lengths, and form the leg notches using your band saw. Use a ½ in. band saw blade to insure straight cuts.

Cut the two rails to length and rabbet the center using your table saw equipped with a dado blade. Carefully mark the area to be cut and make repetitive passes.

The next phase is very important and requires that you have a drill press or accurate drill guide. Carefully mark the top workpieces to be drilled, as well as the top of the legs that fit into the chopping block. Then drill a perpendicular, 5/16 in. diameter hole to accommodate the threaded rod. Also counterbore 1 in. diameter holes into the notched leg areas to accommodate the lag screws that attach the rails. Then pre-drill the holes for the fasteners to allow a ⅜ in. by 4 in. lag screw to pass. Drill two of these holes into each leg. Counterbore 1 in. diameter holes in the outside top pieces to accept the flat washers and nuts.

Next, lay the top on a flat surface that has been covered with plenty of wax paper or newspaper. Lay the top upside down (top down) for assembly. Before gluing, dry-assemble the unit by positioning each layer of the top along with the legs. As you progress, run the threaded rod through each subsequent layer of the top and/or leg. Custom measure the ¼ in. diameter threaded steel rod during the dry assembly and cut to size with a hack saw. As you complete a leg pair, also position the rails in place. When you are satisfied that the assembly is perfect, then disassemble the unit and begin gluing.

Apply a light layer of glue to one joining surface. Then follow this technique for each succeeding layer. Remember to thread the rod through the holes as you progress with each layer. When a leg pair is installed, glue and secure the rails to the legs with a ⅜ in. by 4 in. lag screw and flat washer. After all of the workpieces have been installed, with the exception of the brace, position a flat washer and nut on both ends of the threaded rod and tighten gently. Note that three threaded rods make up the top. Place a flat piece of plywood on top of the ends of the legs to help check for squareness and then square the assembly. When everything is perfect, tighten down the nuts on the threaded rod and on the lag screws. Again, make sure that

This project is courtesy of Georgia-Pacific Corporation.

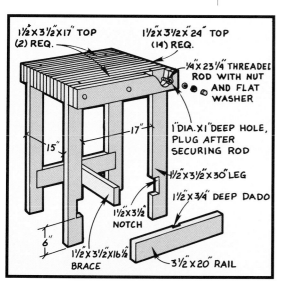

1½ x 3½ x 17" TOP (2) REQ.

1½ x 3½ x 24" TOP (14) REQ.

¼ x 23¼ THREADED ROD WITH NUT AND FLAT WASHER

1" DIA. X 1" DEEP HOLE, PLUG AFTER SECURING ROD

17"

15"

1½ x 3½ x 30" LEG

1½ x ¾" DEEP DADO

1½ x 3½ NOTCH

6"

1½ x 3½ x 16½" BRACE

3½ x 20" RAIL

1 *Make sure the top workpieces to be laminated are absolutely flat. Run them through a thickness planer or dress them on a jointer.*

the table top is sitting flat on the workbench. Allow the glue to cure for at least 24 hours before moving the assembly.

Measure the leg brace and cut to size. Install it between the rails with glue and two ⅜ in. by 4 in. lag screws and flat washers. Install two lag screws at each joint, but do not counterbore into the rails.

Cut 1 in. diameter plugs and fill the counterbored holes. Sand flush with a belt sander.

Finishing

Remove any excess glue that squeezed up during the clamping procedure. Use a paint scraper or an old wood chisel. Then belt sand the top surface along with the edges. Finish sanding the rest of the project with a pad sander, and dull all sharp edges with a sanding block.

Apply your favorite salad oil to the chopping block top and edges. Then rub in a second and third coat after each dries.

Finish the leg portions of the chopping block with a good quality varnish. Sand lightly once the varnish is dried and then apply a second coat to complete the project. — *by Jonathan Wesley.*

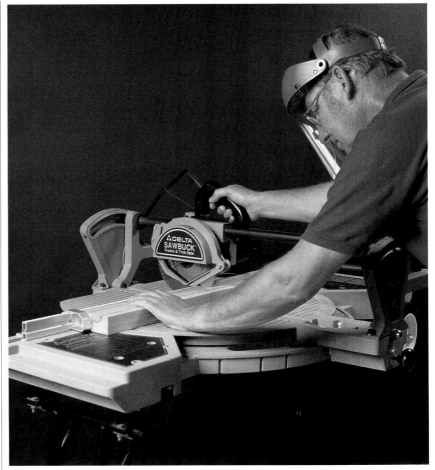

2 *Use a stop block or stop gauge on your saw to cut multiple lengths of workpieces to their proper length. Make sure the blade is cutting square.*

Two Lincoln Clocks

Build one or both of these showpiece clocks.

Tad's pine clock features the full-size drawing below.

Spring-wound clocks were an improvement over weight-driven clocks when they were introduced in the mid-1850s. In 1858, when running for the U.S. Senate, Abraham Lincoln wrote his now classic statement which begins, "You can fool all of the people some of the time..." It may be that as he wrote these words, Lincoln had his eyes fixed on one of the two time-keepers which we offer the wood-worker for reproduction.

For example, he may have been looking at Tad's pine clock, named for Lincoln's son Thomas. Lincoln called him Tad because he thought the child's head looked like a tad-pole.

Tad's pine clock was brought to James Hickey, Curator, Lincoln Collection, Illinois State Historical Library, Springfield, by someone who reported that he was honoring a deathbed promise. The clock apparently had been stolen from Lincoln years before. Today, the clock is in the Illinois Old State Capitol Building in Springfield.

Our second clock, the mahogany Butler clock, is in the Lincoln Home National Historic Site in Springfield. This clock, too, has a story. It belonged to William J. Butler during the period when Lincoln lived in Springfield. Lincoln boarded with the Butlers when he

The mahogany Butler clock.

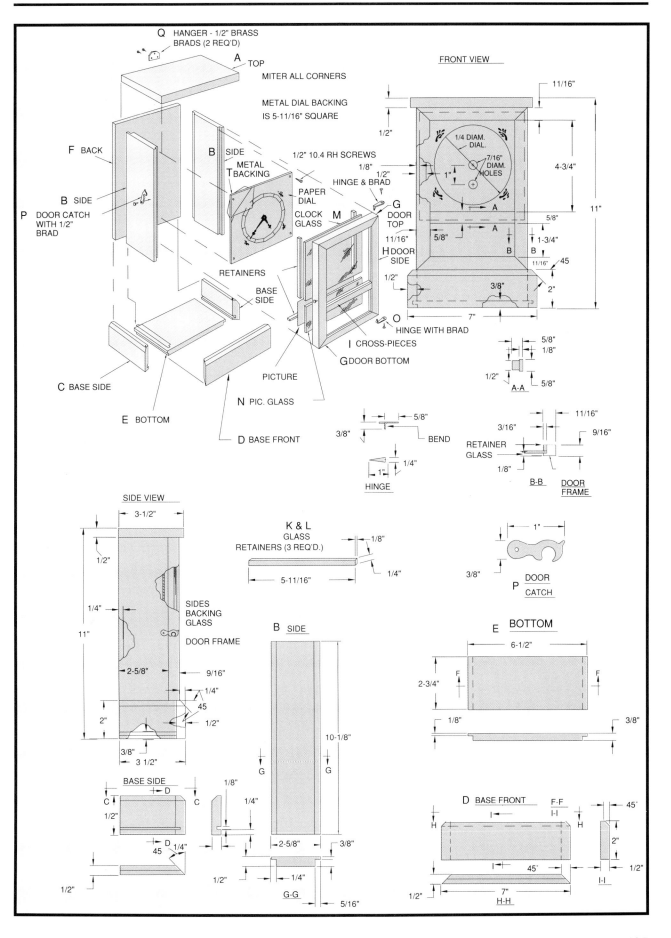

Q HANGER - 1/2" BRASS BRADS (2 REQ'D)

A TOP

MITER ALL CORNERS

METAL DIAL BACKING IS 5-11/16" SQUARE

F BACK

B SIDE

METAL BACKING

1/2" 10.4 RH SCREWS

HINGE & BRAD

PAPER DIAL

CLOCK GLASS

B SIDE

P DOOR CATCH WITH 1/2" BRAD

M

G DOOR TOP

H DOOR SIDE

RETAINERS

BASE SIDE

O HINGE WITH BRAD

I CROSS-PIECES

G DOOR BOTTOM

C BASE SIDE

PICTURE

N PIC. GLASS

E BOTTOM

D BASE FRONT

FRONT VIEW

11/16"

1/2"

1/4 DIAM. DIAL.

7/16" DIAM. HOLES

1"

A

4-3/4"

11"

5/8"

A

5/8"

1-3/4"

1/2"

11/16"

45

3/8"

2"

7"

5/8"

1/8"

1/2"

5/8"

A-A

3/8"

5/8"

BEND

1"

1/4"

HINGE

11/16"

3/16"

9/16"

RETAINER GLASS

1/8"

B-B

DOOR FRAME

K & L
GLASS RETAINERS (3 REQ'D.)

1/8"

1/4"

5-11/16"

1"

3/8"

P DOOR CATCH

SIDE VIEW

3-1/2"

1/2"

1/4"

11"

SIDES BACKING GLASS

DOOR FRAME

2-5/8"

9/16"

1/4"

45

1/2"

2"

3/8"

3 1/2"

BASE SIDE

D

C

C

1/2"

D

45

1/4"

1/2"

1/8"

1/4"

1/2"

1/4"

1/2"

G-G

5/16"

B SIDE

10-1/8"

G

G

2-5/8"

3/8"

E BOTTOM

6-1/2"

F

F

2-3/4"

1/8"

3/8"

D BASE FRONT

F-F

I-I

H

H

I

I

45°

7"

H-H

1/2"

45°

2"

1/2"

I-I

127

*T*he paper clock face is glued to the metal backing plate. The plate is fastened into rabbets in the case with small screws at the corners.

*G*lue and clamp the mitered door frame workpieces. When the joints are dry, bore pilot holes in the corners and reinforce them with 3d nails.

BILL OF MATERIALS — Tad's Pine Clock

Finished Dimensions in Inches

A	Top	½ x 3½ x 6⅝ pine	1
B	Side	½ x 2⅝ x 10⅛ pine	2
C	Base Side	½ x 2 x 3½ pine	2
D	Base Front	½ x 2 x 7 pine	1
E	Bottom	⅜ x 2¾ x 6½ pine	1
F	Back	¼ x 5½ x 10½ pine	1
G	Door Top/Bottom	⁹⁄₁₆ x ¹¹⁄₁₆ x 6 pine	2
H	Door Side	⁹⁄₁₆ x ¹¹⁄₁₆ x 10⅛ pine	2
I	Crosspiece	⁷⁄₁₆ x ½ x 5 pine	1
J	Crosspiece	⅛ x ⅝ x 4⅝ pine	1
K	Glass Retainer	⅛ x ¼ x 4¹⁵⁄₁₆	2
L	Glass Retainer	⅛ x ¼ x 4¹⁵⁄₁₆	1
M	Glass	4⁵⁄₁₆ x 4⁵⁄₁₆ single-strength glass (for clock)	1
N	Glass	1⁵⁄₁₆ x 4⁵⁄₁₆ single-strength glass (for picture)	1
O	Hinge	¼ x 1 20-gauge brass	2
P	Door Catch	1 x 2 20-gauge brass	1
Q	Hanger	⅝ x ¾ 20-gauge brass	1
R	Picture	2 x 5 (see note below)	1
S	Paper Clock Dial	4¼ dia. dial with ⅝ Roman numeral (see note below)	1
T	Metal Backing	5¹¹⁄₁₆ x 5¹¹⁄₁₆ 20-gauge sheet metal (for clock dial)	1
U	Clock Movement	Urgo 8-day pendulum clock movement (see note below)	1
V	Clock Hands	1⅜ and 2⅛ serpentine clock hands (see note below)	2

started his law practice. The mahogany Butler clock now rests in the sitting room of the Lincoln Home National Historic Site.

Neither timepiece shows a clockmaker's name. For these reproductions, Urgos pendulum clock movements from West Germany are used. While the originals are 3-hour mechanisms, these are 8-day clocks. There are other minor differences. The mechanisms of the original clocks are screwed to the back of the case, but the mechanisms for these reproduction clocks mount on a sheet metal face. The winding arbor on these clocks is at six o'clock, it is at eleven and three on the originals. And finally, the pendulum drop on these clock plans is about ¼ in. shorter than in the original Lincoln clocks.

Tad's Pine Clock

Start Tad's pine clock by ripping 18 in. of base stock and cutting a ¼ in. chamfer. Then cut the base stock into two equal lengths. Cut a ⅛ in. by ¼ in. groove in one piece for the bottom and miter for the sides of the base.

Next, rip two pieces of ½ in. stock for the two sides of the clock. Cut rabbets in the sides, one for the face of the clock, the other for the back, using an overhead router, saw or jointer. Cut out the bottom, and

cut in a ⅛ in. by ¼ in. rabbet. Then, making certain the top edges of the base are even, glue and 3d nail the base and bottom in so as to square the base; clamp if necessary.

Sand all exposed edges of the clock case pieces. When the glue on the base and bottom assembly has dried, glue and clamp the sides to the base, making sure the sides are square. Cut the back to fit, then glue and 3d nail in place. Next, cut the top to size and affix, using the same procedure.

To construct the door, rip 33 in. of ¹¹⁄₁₆ in. stock, ⁹⁄₁₆ in. wide. Cut in

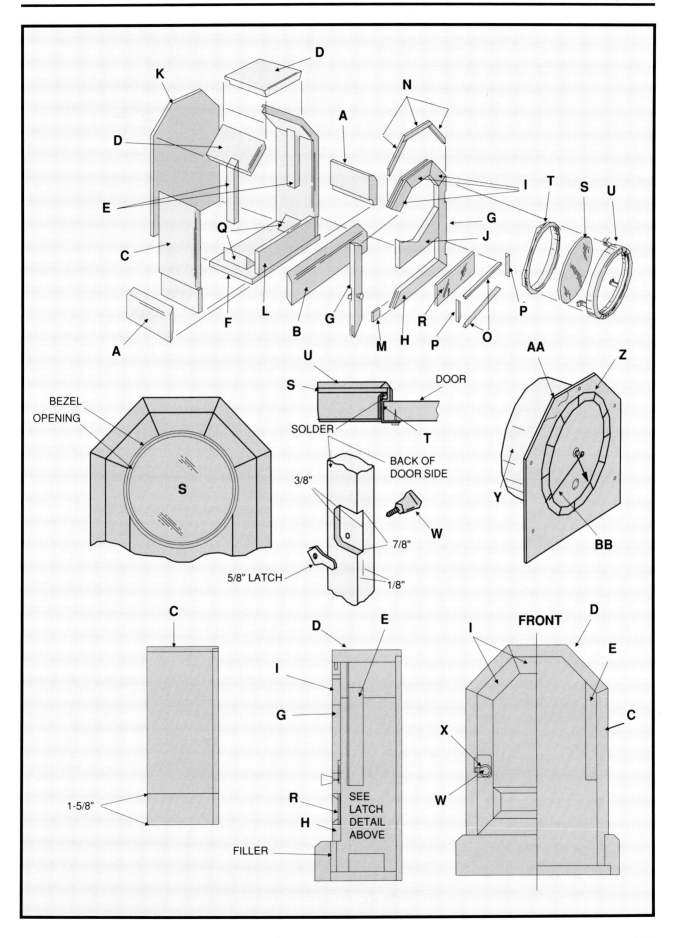

BEZEL
OPENING

S

SOLDER

DOOR

BACK OF
DOOR SIDE

3/8"

7/8"

5/8" LATCH

1/8"

W

C

1-5/8"

D

I

G

R

H

E

SEE
LATCH
DETAIL
ABOVE

FILLER

I

FRONT

D

E

C

X

W

AA

Z

Y

BB

a rabbet for single-strength glass, and sand the inside edges with 100 grit sandpaper. Glue up the door. When the door is dry, cut cross pieces to correct lengths and glue into place on the door. After the glue has set, 3d nails may be added to give the structure extra strength. However, bore pilot holes to prevent splitting the wood.

Sand all remaining edges of the door with 100 grit sandpaper. Sand upper and lower door edges at a 2 or 3 degree inward angle. This will allow the door to close smoothly. When properly sanded, the front of the door will be about ⅛ in. longer than the back of the door.

Sand back, bottom and sides flush, and break all corners. Stain the clock case to the desired color. We used dark walnut made by Minwax. Allow the stain to dry thoroughly, then rub on several coats of French polish, consisting of 80 percent orange shellac and 20 percent boiled linseed oil. When this has dried, buff with carnuba wax.

The door is mounted on pin hinges, which can be made from 20-gauge brass. Cut two pin hinges 1 in. by ¼ inch. Drill a hole for a small brass nail. File the hinges smooth, and bend to the specifications on the drawing. Next, bore a hole, centered at the top and bottom of the door, ⅜ in. in from the corner. Use a No. 60 drill bit. When mounting the hinges on the door, bend the lower hinge upward slightly, so that the door does not rub on the base.

Then cut the paper face to fit the opening, and cut the sheet metal to fit the face. Drill holes for the hands, winding arbors and mounting screws. Apply the face to the metal with rubber cement. Paint flowers at four corners of the dial with light green paint to match the decorative scene. Use a medium brush to get the tear-drop shapes of petals. File the paper away from the arbor holes with a round file. Mount the movements to the face, and screw the assembly into the case.

Cut two strips to hold the top and one strip to hold the bottom glass pieces in place. Glue the ¼ in. by ½ in. strips and glass with epoxy. Complete the clock by cutting the

BILL OF MATERIALS — Mahogany Butler Clock

Finished Dimensions in Inches

A	Base Side	⅝ x 1⅝ mahogany	2
B	Base Front	⅝ x 1⅝ x 7 mahogany	1
C	Side	½ x 3 x 7¼ mahogany	2
D	Top	½ x 3 x 2½ mahogany	3
E	Mounting Block	½ x ¾ x 4 mahogany	2
F	Bottom	⅜ x 2¾ x 6¼ mahogany	1
G	Door Side	⅜ x 11/16 x 5¾ mahogany	2
H	Door Bottom	⅜ x 11/16 x 5 mahogany	1
I	Door Top	⅜ x 11/16 x 2 1/16 mahogany	3
J	Door Panel	⅜ x 2 x 3⅝ mahogany	1
K	Back	¼ x 5½ x 9⅛ mahogany	1
L	Filler	¼ x 1¼ x 5 mahogany	1
M	Spline	⅛ x ¾ x ¾ mahogany	2
N	Spline	⅛ x 5/16 x 3 (trim to fit)	3
O	Moulding	3⅝ quarter-round moulding for top/bottom (cut quarter rounds from ½ dia. dowel x 6)	2
P	Moulding	1¼ quarter-round moulding for sides (cut from dowel as above)	2
Q	Glue Block	¾ x ¾ x 1½	2
R	Mirror	⅛ x 1¼ x 3⅝	1
S	Glass	3 7/16 dia. single-strength glass (for clock)	1
T	Bezel	4 dia. brass clock bezel	1
U	Retainer	¼ x ⅝ 20-gauge brass	3
V	Hardware	½ x ¾ brass butt hinge	2
W	Hardware	½ x ½ brass knob	1
X	Door Latch	5/16 x ⅝ 20-gauge brass	1
Y	Movement	Urgo 8-day pendulum clock movement	1
Z	Dial Face	3½ dia. paper dial face	1
AA	Metal Backing	5 x 5 20-gauge sheet metal (metal backing for clock dial)	1
BB	Clock Hands	1½ and 1¾ spade clock hands	1

steam engine scene to fit tightly, and press into place between the strip and the glass.

Make the door catch out of 20-gauge brass, and mount to the clock side with ⅜ in. long brass brads into the case. Nail another brass brad to the door to receive the catch.

Use marks on the top of the case indicate that this clock was hung on the wall. Snip the clock hanger piece out of brass stock. Drill holes for the brads and the holding hole. Fasten with brads to the back of the case.

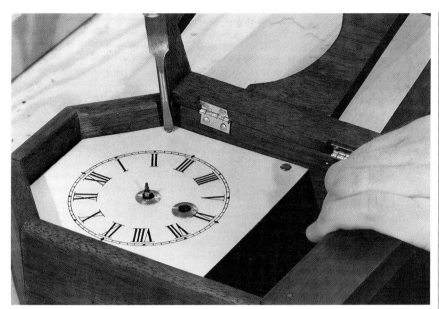

To make the Butler clock face, first cut paper clock face to fit the case opening. Then use the paper cutout to transfer the shape onto sheet metal. Cut and bore the metal and glue the paper to it. The movement mounts on the back of the metal with brass nuts.

Mahogany Butler Clock

To begin construction of the mahogany clock, study the drawing carefully.

Rip 24 in. of side and top stock. Mark 1¼ in. at each end of the stock, and round off the edge of the portion between the marks. Use a spokeshave and chisel to round the edge. Then cut the sides and top pieces to length. Next, notch out the lower outside edge of the side pieces to receive the face.

Cut the bottom to size. Rip the base stock 18 in. long by ⅝ in. thick. With the spokeshave or plane, round one edge of the base stock. Sand and cut the stock into two equal lengths. Cut the front of the base to size from one length. Cut a rabbet for the bottom in the other piece, and miter all of these pieces to correct lengths.

Next, glue the sides of the case to the sides of the base. When dry, glue and 3d nail the front of the base, and glue the bottom in. Drill pilot holes to prevent splitting the stock. Next, miter the three top pieces per the drawing; glue and 3d nail into place. Finally, cut the back piece to fit, and glue and nail it to the rear of the case.

To construct the door, rip 30 in. of stock, and cut the pieces to size; trim angle with a miter trimmer.

Then saw a slot ⁵⁄₁₆ in. deep for splines, and cut the splines ⅜ in. to fit tightly into the grooves. Notch out for the door catch on the left side piece. Then cut the panel to size above the mirror opening.

Dry-assemble the door, making sure that it fits the opening. Then glue the door pieces together. When dry, sand the face and edges smooth with 120 grit sandpaper. Sand the outside edges of the door with a slight inward angle of about 2 or 3 degrees, so that the door closes properly as it recesses into the case.

Next, glue a ¼ in. filler strip on the inside of the front face, and glue ½ in. stock for mounting the face.

Make the small quarter-round in front of the mirror by ripping a ½ in. dowel in half; then rip the halves in half, and miter to length. Next, paint the small quarter-round with gilding.

To finish the clock case, dip it into a solution of one tablespoon lye per gallon of water. The solution bleaches the wood. (Use caution when using lye; wear gloves and goggles.) When the desired shade is reached, wash the case and door with water and then wipe dry. Finally, apply French polish. Fill nail holes with an appropriate shade of wax stick. Buff with carnuba wax.

Cut a catch from 20-gauge brass,

tap metal to fit the screw on the knob and assemble with glue. Cut a slot in the clock case for the catch, using a Dremel saw attachment or knife.

Next, cut the paper face to fit the opening, then cut the sheet metal to fit the face. Drill holes in the metal for hands, winding arbors and for No. 4 by ½ in. mounting screws. Apply a thin, even coat of rubber cement to the metal and affix the paper face. File paper out from arbor holes with a round file. Then fasten the clock movement to the face with nuts provided and screw the face to the case.

Then mortise hinges in the door; screw the hinges on the door, then onto the clock body. Glue quarter-round to the door and mirror with epoxy. Solder the brass pieces to hold the glass in bezel and bezel to frame. Fasten the bezel to the door. Finally, put on the hands and pendulum.

You now have two fine clock reproductions straight out of Lincoln history that you can display proudly. As a clock buff, you will find the truth Lincoln had in mind when he said, "People who like this sort of thing will find this the sort of thing they like."

Full-size plans are available for $9.95 each (specify the clock) from Creative Features, Suite 1317, Six North Michigan Ave., Chicago, IL 60602. The paper clock face is also available from this source for $2.50. The mahogany was obtained from Craftsman Wood Service, 1735 West Cortland Court, Addison, Illinois 60101. — *by David Warren. Photographs by Matt Doherty.*

Sources:

Constantine, 2050 Eastchester Rd., Bronx, NY 10461 (kits, parts and fine hardwoods).

Craftsman Wood Service, 1735 West Cortland Ct., Addison, IL 60101 (kits, parts and fine hardwoods).

Crown Clock Co., 756 Nicholas Ave., Drawer G, Fairhope, AL 36532 (kits and parts).

Emperor Clock Co., Industrial Park, Fairhope, AL 36532 (clock kits).

Mason & Sullivan Co., 39 Blossom Ave., Osterville, MA 02655 (kits and parts).

Barbecue Cart

Cooking outdoors is easy with this portable barbecue cart.

There is nothing like having everything you need at hand for a backyard barbecue. This barbecue cart features a tile countertop, a rack for hanging cooking utensils and roomy storage shelves. Attach casters and the unit can be easily rolled inside to keep it out of the rain.

Tips

Because the cart will be rolled around on both smooth and rough surfaces, it is best to use wood screws and carpenter's glue for securing the project's main joints. Make sure to use noncorrosive hardware, and sink all nail heads. Fill in the nail holes with a suitable colored wood filler.

Construction

Begin by cutting to length the top and bottom rails using a radial arm saw. Keep the good side of the wood up when making the cut on a radial arm saw. This prevents visible wood tear-out on the final project.

Next, cut the four legs to length on your radial arm saw. Secure the longest top and bottom rails to two leg pairs with No. 8 by 1¼ in. flathead wood screws (countersunk) and carpenter's glue. Use a carpenter's square to square the assembly.

Similarly, assemble and secure the other two long top and bottom rails to the remaining two leg pairs. Match up the side assemblies to insure they are symmetrical.

After the frame assembly has dried, secure the short top and bottom rails to the frame assembly with No. 8 by 2 in. flathead wood screws (countersunk) and carpenter's glue.

Rip and crosscut the ¾ in. plywood top, two shelves and side panel to their proper sizes. Use a circular saw, guided by a straightedge to insure straight cuts. Don't forget to use a plywood circular saw blade for a smooth cut. Attach the bottom shelf with glue and 4d finishing nails.

Now crosscut four shelf cleats to size, and secure them to the legs with No. 8 by 1¼ in. flathead wood screws (countersunk) and carpenter's glue. Install the upper shelf with carpenter's glue and 4d finishing nails. Similarly secure the top planks.

Turn the cart upside down and cut the two axle supports as well as the footrest. Then drill ⅞ in. diameter holes into the axle supports so that the axles clear the bottom rails. Install both axle supports and the footrest with No. 8 by 1¼ in. flathead wood screws (countersunk) and carpenter's glue. Both axle supports must be located symmetrically on the frame in order for the axle to turn freely.

Form the wheels by drawing a 4 in. diameter circle on ¾ in. material. Then cut out the perimeter on a band saw equipped with a ⅜ in. blade. Complete the wheel by drilling a 1³⁄₁₆ in. diameter hole with a drill press at the wheel's center. Follow the same procedures in completing the second wheel.

Cut the axle to length and run it through the axle supports. Then glue the two wheels onto the axle with carpenter's glue.

Now cut the two handles to their proper lengths and round the ends on a stationary disk sander or by cutting them on a band saw. Then drill a ¾ in. diameter hole with a drill press.

Position the handles on the upper legs and clamp them in place with C-clamps. Install them at a slight 15 degree angle, making sure

This project is courtesy of Georgia-Pacific Corporation.

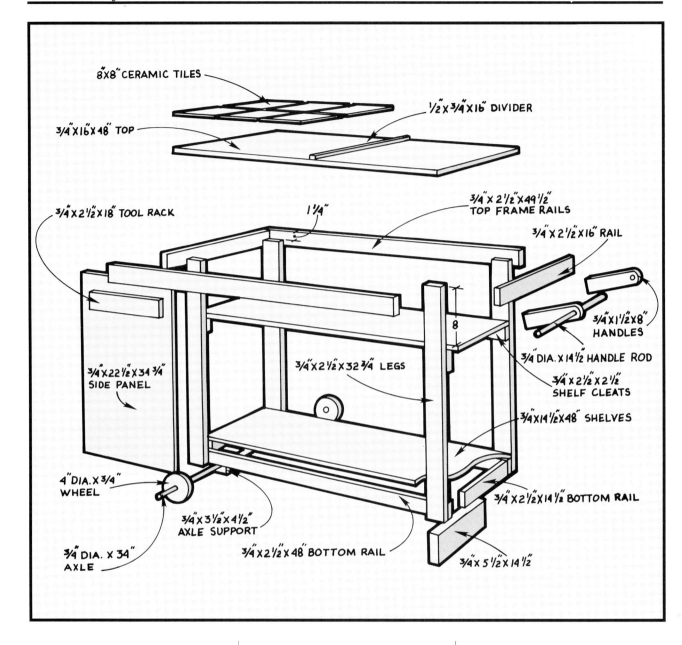

8"X8" CERAMIC TILES

1/2" X 3/4" X 16" DIVIDER

3/4" X 16" X 48" TOP

3/4" X 2 1/2" X 49 1/2" TOP FRAME RAILS

3/4" X 2 1/2" X 16" RAIL

3/4" X 2 1/2" X 18" TOOL RACK

1 1/4"

3/4" X 1 1/2" X 8" HANDLES

3/4" DIA. X 14 1/2" HANDLE ROD

3/4" X 2 1/2" X 2 1/2" SHELF CLEATS

8

3/4" X 22 1/2" X 34 3/4" SIDE PANEL

3/4" X 2 1/2" X 32 3/4" LEGS

3/4" X 14 1/2" X 48" SHELVES

4" DIA. X 3/4" WHEEL

3/4" X 3 1/2" X 4 1/2" AXLE SUPPORT

3/4" DIA. X 34" AXLE

3/4" X 2 1/2" X 48" BOTTOM RAIL

3/4" X 2 1/2" X 14 1/2" BOTTOM RAIL

3/4" X 5 1/2" X 14 1/2"

that both handles are located at the proper height. Drill holes for the 3/8 in. stove bolts which will be used to secure each handle. Drill from the outside leg to eliminate visible wood tearout. Then install the 3/8 in. by 2 in. long stove bolt with flat washer and nut.

Install the 3/4 in. diameter handle with carpenter's glue.

Make your own tool rack or buy a supplied one. If you purchase one, do not cut the tool rack, but simply attach the purchased rack to the side panel. If you opt to make your tool rack, drill 1/2 in. diameter holes into the tool rack at a slight 15 degree angle. Make sure that you have enough holes to accommodate barbecue items and space these holes accordingly. Cut 1/2 in. doweling to 2 1/4 in. lengths, and round the end of each length. Then glue the doweling onto the racks.

Mount the tool rack to the side panel with No. 8 by 1 1/4 in. flathead wood screws (countersunk) and carpenter's glue. Then attach both side panels to the barbecue cart assembly with No. 8 by 1 1/4 in. flathead wood screws (countersunk) and carpenter's glue.

Install the ceramic tiles using a proper adhesive for exterior use. Then cut and install a 3/4 in. divider with glue and 1 in. brads.

Finishing

Remove the stove bolts driven into the handles and finish sand the entire project, making sure that all sharp corners are slightly rounded so that the wood is suitable for accepting a finish.

Apply your favorite stain, being careful to work it into the wood. Then apply a good quality exterior polyurethane finish. Remember to finish all wood surfaces, even under the shelf areas. After it has dried, lightly sand the project and apply a second coat. Now add casters to complete your project. — *by Jonathan Wesley.*

Lincoln's Armoire

You can build a wardrobe cabinet like the one Lincoln used. It is a practical adaptation of an irreplaceable treasure.

Here we show you how to build a wardrobe, or armoire, which Lincoln used as an adult in Springfield, Illinois.

Closets in bedrooms are a modern concept. Lincoln's Springfield home had only a trunk room for storage. In place of a closet, Lincoln used this wardrobe for his black wool suits, stovepipe hat, boots and other articles of clothing. His armoire was supplemented by a mahogany chest of drawers.

Our adaptation of Lincoln's armoire can serve the same function. We've provided plenty of adjustable shelves on which to store shirts, sweaters and other items for ready access.

The original Lincoln wardrobe stands in the Lincoln Home National Historic Site in Springfield. Constructed of light-colored walnut, its maker is unknown.

An unusual feature of the cabinet is that it readily disassembles for transfer from room to room. Our adaptation uses glued-up pine and plywood for economy and strength. We omitted the knockdown feature to keep construction simple.

In 1861, when he left for Washington, Lincoln sold his wardrobe for $20, or about two weeks' wages for a journeyman.

Today, a quality pine armoire sells for $1,250 to $2,000. This do-it-yourself wardrobe can cut costs by $1,500.

Construction

To begin, glue up enough pine boards to make the sides, doors, six shelves and interior partition of the wardrobe. Since this adaptation departs from time-honored rail-and-stile joinery paneled construction, it is important that seasoned, dry wood be used to avoid warps and twists after construction.

Mark facing edges to be jointed, and carefully plane each edge straight for a smooth, tight fit. Then apply glue to both edges. Clamp with equal pressure, and use a straightedge or carpenter's square to insure that the assembled boards are flat.

While the glue is drying, lay out the top and bottom on ¾ in. plywood, sanded on one side. Lay out the back on ½ in. plywood, sanded on one side. Plan these components so the good face (sanded side) is exposed, but draw on the back side. To minimize tearing when these pieces are cut out with a circular saw, cut from the back side. Cut to the outside of the lines so your dimensions will remain accurate.

Unclamp the sides and partition boards, then sand, carefully smoothing the glued joints so that they will accept stain evenly. Then lay out each component, and use a circular saw to cut them to size.

Assemble the box of the wardrobe face-down on a flat surface. Mark the location of the bottom. Then glue and nail the sides to the bottom. Locate and fit the partition, and glue and nail this to the bottom. Finally, position the top, and glue and nail, making sure to correctly position the partition.

Next, draw a line on the rear face of the back panel showing the center of the partition. The assembly will square up when the back is glued and nailed in place. Remember to keep

This 19th-century walnut armoire was used daily by Abraham Lincoln during his years in Springfield, Illinois. It is preserved in his bedroom at the Lincoln Home National Historic Site in Springfield. (Measurements and photos by special permission of the U.S. Department of the Interior, National Park Service.)

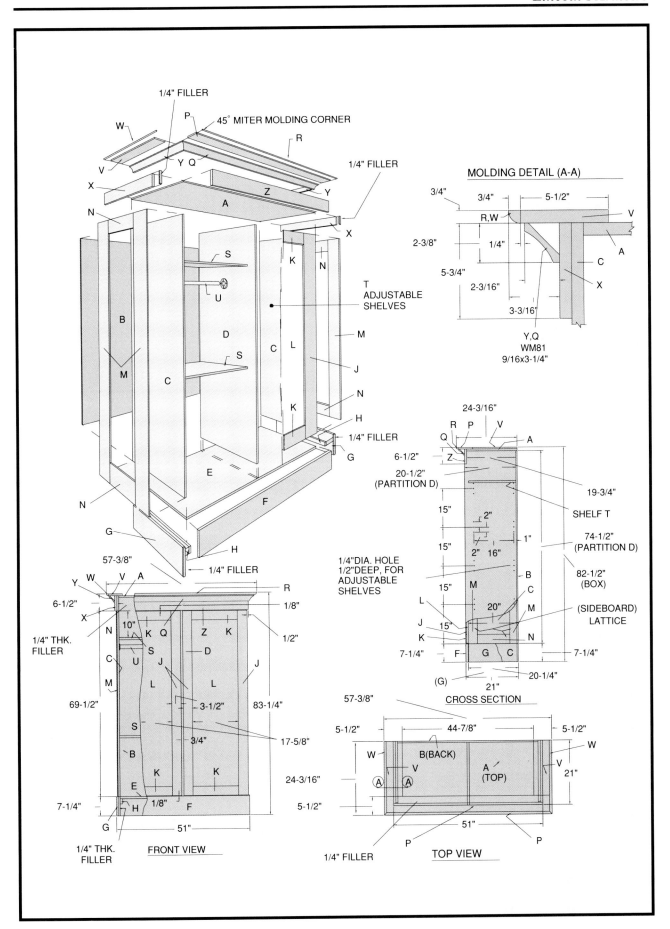

1/4" FILLER

P

45° MITER MOLDING CORNER

W

R

V

Y Q

X

1/4" FILLER

Z

Y

A

N

S

K

N

U

T
ADJUSTABLE
SHELVES

B

D

C

L

M

S

C

J

M

N

K

H

1/4" FILLER

G

E

F

N

G

H

57-3/8"

1/4" FILLER

MOLDING DETAIL (A-A)

3/4"

3/4"

5-1/2"

R,W

V

2-3/8"

1/4"

A

5-3/4"

C

2-3/16"

X

3-3/16"

Y,Q
WM81
9/16x3-1/4"

24-3/16"

R P V

Q

A

Z

6-1/2"

20-1/2"
(PARTITION D)

19-3/4"

15"

SHELF T

2"

15"

1"

74-1/2"
(PARTITION D)

2" 16"

B

82-1/2"
(BOX)

15"

(SIDEBOARD)
LATTICE

L

C

J

20"

M

15"

K

N

7-1/4"

F G C

7-1/4"

1/4"DIA. HOLE
1/2"DEEP, FOR
ADJUSTABLE
SHELVES

M

(G)

20-1/4"

21"

57-3/8"

CROSS SECTION

5-1/2"

44-7/8"

5-1/2"

W

B(BACK)

W

V

A
(TOP)

V

21"

A A

Y W V A

6-1/2"

R

X

1/8"

10"

K Q

Z K

1/2"

1/4" THK.
FILLER

N

C

U

S

J

D

J

M

L

L

J

69-1/2"

3-1/2"

83-1/4"

S

17-5/8"

3/4"

24-3/16"

B

K

K

E

5-1/2"

7-1/4"

1/8"

F

H

G

51"

P

P

1/4" THK.
FILLER

FRONT VIEW

1/4" FILLER

TOP VIEW

1 *To prepare pine boards for gluing, carefully plane edges true and square. Mark mating edges before gluing.*

2 *Cut the back to length with a hand saw. Cut with the good face up.*

the less-perfect sides of the plywood components hidden from view. Place nails every 8 to 10 in. to secure all glued components. Use the same separation when nailing through the back into the partition and through the sides into the back.

When the glue has dried, sand the assembly. With the cabinet on its back, cut and apply the base and top trim pieces. Turn the cabinet with one side up, and apply base and top trim. Also, cut then glue and nail the cleat. Follow by applying lattice, per the drawing, with glue and ¾ in. brads, or by clamping. Repeat on the opposite side.

Next, make a shelf clip template guide per the drawing. Stand the cabinet erect, and locate and bore shallow ¼ in. holes down the inside surface of the left side, near the front and back edges. Repeat this procedure on the left inside surface of the partition.

Then lay out and saw to size each of the six shelves. Note that the fixed shelves for the right-hand section of the wardrobe are ½ in. wider than the adjustable shelves. After sawing with a circular saw, plane or sand the edges smooth. Test-fit each shelf. Finally, locate fixed shelves into the right-hand section, then glue and nail.

Next, use a miter box to saw the cove pieces for the top edge of the cabinet. Sand each end of the cove pieces carefully and test-fit to insure a tight miter joint. Then glue and nail

BILL OF MATERIALS — Lincoln's Armoire

Finished Dimensions in Inches

A	Top	¾ x 20¼ x 48 A/C plywood	1
B	Back	½ x 48 x 81¾ A/C plywood	1
C	Side	¾ x 20 x 82½ pine	2
D	Partition	¾ x 20½ x 74½ pine	1
E	Bottom	¾ x 19¾ x 48 A/C plywood	1
F	Front Base	¾ x 7¼ x 51 pine	1
G	Base Side	¾ x 7¼ x 20¼ pine	2
H	Side Cleat	¾ x 2 x 19¾ pine	2
J	Vertical Door Trim	¼ x 3½ x 69¼ lattice	4
K	Top/Bottom Door Trim	¼ x 3½ x 18⅜ lattice	4
L	Door	¾ x 24⅜ x 69¼ pine	2
M	Vertical Side Trim	¼ x 3½ x 69½ lattice	4
N	Top/Bottom Side Trim	¼ x 3½ x 13 lattice	4
P	Top Trim Front	¾ x 5½ x 55⅞ lattice	1
Q	Top Front	3¼ x 55⅜ cove	1
R	Edge of Top Trim	¾ x 57⅜ quarter round	1
S	Fixed Shelf	¾ x 19 x 23⅝ pine	2
T	Adjustable Shelf	¾ x 19 x 23⅜ pine	4
U	Clothes Bar	1⁵⁄₁₆ dia. x 23⅜ dowel	1
V	Top Trim at Side	¼ x 5½ x 17¹⁵⁄₁₆ lattice	2
W	Top Side Trim Edges	¾ x 24³⁄₁₆ quarter round	2
X	Side Fascia	¾ x 5¾ x 20¼ pine	2
Y	Top Side	3¼ x 23³⁄₁₆ cove	2
Z	Fascia	¾ x 5¾ x 51 pine	1

3 *Glue edges and clamp with pipe clamps, applying pressure equally. Test flatness of assembly with a straightedge.*

4 *Cut lattice trim to size. Glue and nail to the sides. The trim is fitted tightly between top and base trim.*

5 *Miter 3¼ in. cove pieces to length and sand and fit edges. Then glue and nail the cove pieces with finishing nails to the top of the cabinet.*

6 *Any dents can be raised by steaming them with a damp cloth and a soldering iron with special flat tip.*

these pieces to the top of the cabinet, using 6d finishing nails. Sink nails. Attach top trim. Complete by mitering quarter rounds and gluing and nailing them on.

Complete the doors and attach lattice per the drawing. Check the doors carefully. Make sure they are straight, without warp or twist. Lay them on a flat surface when gluing on lattice. This will insure a true finished product. Then locate and mortise out for the hinges. Also mortise for the hinges on the cabinet frame.

Complete the Lincoln armoire by attaching hinges to the doors and mounting them to the cabinet. Also, cut a clothes bar from a 1⁵⁄₁₆ in.

dowel, mount hanger hardware and position the bar. Then insert shelf clips at desired locations and fit in the adjustable shelves. If desired, secure clips to the shelves with No. 5 by ½ in. flathead brass screws.

Finishing

Fill holes and steam out dents. Then disassemble the hardware. Carefully sand the entire assembly and each shelf board using 120 grit and 180 grit paper. After sanding, wipe down thoroughly with a tack cloth to remove all dust.

Brush on two thin coats of Minwax pecan polyshade, which combines a light stain and polyurethane. Allow overnight drying between

coats. Reinstall shelves and doors. To achieve a soft, satin finish, rub all surfaces lightly with fine steel wool and wax.

You now have a handsome pine wardrobe that will serve as a handy place for everyday or seasonal clothing storage. And at the same time, you'll have a sense of history for the time when Lincoln lived in Springfield.

Large-format plans, which also include the original Lincoln armoire, are available for $16.95 from Creative Features, Armoire, Suite 1317, Six North Michigan Ave., Chicago, IL 60602. — *by David Warren. Photography by William Laskey. Photo styling by Virginia Howley.*

Redwood Planter

Liven up your deck, patio or garden with this handsome redwood planter.

This planter is easy to make and constructed of long-lasting, weather-resistant redwood. At 30 in. square, the planter is large enough for bedding plants or tropical trees. For an original gift idea, you might want to build several planters.

Tips

Use Construction Heart redwood for any pieces that may come in contact with the soil. For best results,

This project is courtesy of the California Redwood Association.

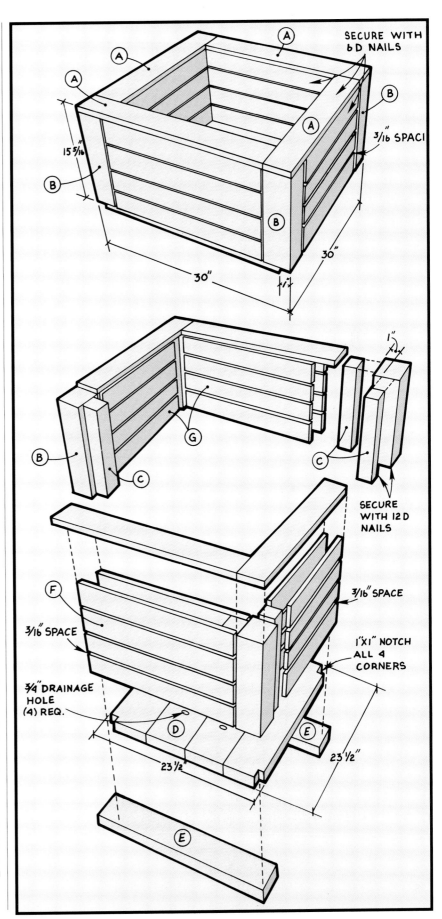

SECURE WITH 6 D NAILS

3/16" SPACI

15 5/16"

30"

30"

SECURE WITH 12 D NAILS

3/16" SPACE

3/16" SPACE

1'X1" NOTCH ALL 4 CORNERS

3/4" DRAINAGE HOLE (4) REQ.

23 1/2"

23 1/2"

1"

measure and trim each piece as you proceed. Make sure to pre-drill holes, especially at the ends of boards, to prevent splitting. Use noncorrosive nails and hardware such as stainless steel, aluminum or hot-dipped galvanized.

Construction

Begin by cutting the corners (B) and cleats (C) on a radial arm saw or power miter box. Then center and nail two cleats to each corner workpiece with 12d nails.

Now cut the outer sides (F) to the proper lengths and secure them to the cleats. The top and bottom sides are flush with the outside of the corner. Remember to keep a $3/16$ in. space between the end of each side and the corner. Refer to the illustration. The two remaining side workpieces are spaced evenly between the ones that you just installed.

Now cut the inner sides (G) to suit the inside of the redwood planter. Note that these also have $3/16$ in. spaces between boards. However, they butt against the corners (B).

Now attach the four workpieces for the top (A) to the planter with 6d nails. Cut these pieces to suit and make sure that there are no gaps where they adjoin one another.

Now cut the two pieces for the base (E) and secure to the bottom of the planter. Inset each base 1 in. from the planter's outer side. Then custom fit four workpieces for the bottom (D) to suit the inside base of the planter. Make sure to cut 1 in. by 1 in. notches in all four corners where they adjoin the corner pieces.

It is not necessary to dowel the bottom unit. Secure the bottom to the two bases with 6d nails and then drill four $3/4$ in. diameter holes into the assembled bottom. This will allow adequate water drainage.

It is a good idea to line the inside of the planter with a polyethylene liner, drilling holes in the bottom of the liner for drainage.

Finishing

As with all redwood projects, you can either let the 4 x 4 planter weather naturally or apply a suitable

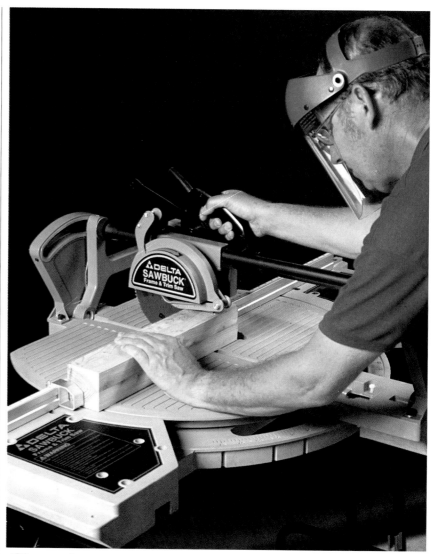

*C*rosscut the corners (B) to length on your power saw. Check to make sure the tool is cutting square. Wear safety goggles.

finish, such as a water repellent. Make sure not to use any mildewcides or other finishes containing toxic materials or chemicals, since these may kill the plants that you wish to have flourish. — *by Jonathan Wesley; design by Environmental Planning & Research; photograph by F. B. Stimson.*

BILL OF MATERIALS —Redwood Planter

Finished Dimensions in Inches

A	Top	$3/4$ x $3\frac{1}{2}$ x $26\frac{1}{2}$ redwood	4
B	Corner	$3\frac{1}{2}$ x $3\frac{1}{2}$ x $14\frac{9}{16}$ redwood	4
C	Cleat	$1\frac{1}{2}$ x $1\frac{1}{2}$ x $14\frac{9}{16}$ redwood	4
D	Bottom	$1\frac{1}{2}$ x $7\frac{1}{4}$ x 25 redwood	4
E	Base	$1\frac{1}{2}$ x $3\frac{1}{2}$ x 28 redwood	2
F	Outer Side	$3/4$ x $3\frac{1}{2}$ x $22\frac{5}{8}$ redwood	16
G	Inner Side	$3/4$ x $3\frac{1}{2}$ x 25 redwood	16

Victorian Fireplace Bench

Sit by the fireplace or wood stove on this cozy bench.

This replica of an actual Victorian fireplace bench is constructed from pine and comfortably seats two people.

Construction

Using a template, lay out the shape of the four feet on clear 4 x 4 stock. Cut out the shapes on a band saw. Sand with drum and disk sanders, using 120 grit paper. Then make rabbet cuts where indicated in the drawing.

To make the front and rear panels, joint and glue together a 1 x 12 and 1 x 6 board. Then rip the panels to a 15 in. width. Finally, cut the curve on the top of each piece with a band saw.

Make all four corner uprights simultaneously by stacking four boards together (see photo). Lay out the design and cut with a band saw, except for the lower portion. Cut the lower portion of the corner uprights on a table saw, being careful not to cut up into the curved area. Finish the cut by hand, and clean it with a chisel or wood rasp.

Sand the corner pieces while they are still stacked, using a disk sander for concave shapes. Use a fine rasp to smooth tight spots, such as where the curve meets the straight-cut portion. When finished, separate the stack, and rout ¼ in. grooves on each outside face per the drawing. Later, these grooves receive beading.

The corner uprights also have a routed groove to accept the pillow rails. Make a template of this area, and transfer its contour onto all four uprights. Then rout, freehand, up to the contour line; use wood chisels to complete the recess.

Glue and dowel three 1 x 6s to form the side panels. Next, cut the notch and rabbet that allows each end panel to lock into the corner upright. Determine where the notch is located by positioning the upright and end panels next to each other.

Make an end template for shaping the pillow rails. It should fit the routed area that was cut into the uprights, yet allow room for inserting the end panels. Cut the clear 2 x 4 stock to length for both rails. Use a jack plane to cut convex areas, or

a round or gutter plane for concave portions. Then sand until smooth. Use a tilting arbor table saw to cut the rabbet for the end panels in the full length of the rails.

Drill the dowel holes into the rail and end panels. Make a V-shaped jig to hold the rail for boring the dowel holes. Next, join the rail and the end panel with glue and dowels.

Bore the remaining dowel holes for assembling the end panels, uprights, and front and rear panels. Dry-fit the end panels with the glued rail to the corner uprights. Finally, glue and clamp these pieces together.

Next, cut two 1 x 3 rails to fit inside the front and rear panels. Make sure each rail is parallel, and bore ½ in. diameter holes with a spur bit, spaced to match the mattress plans. These holes are bored all the way through the rails and no more than ½ in. into the rear panels. Clamp the rails to the panels with glue.

When dry, assemble the front and rear panels to the assembled uprights with dowels and glue. Square the unit before the glue sets up.

Cut the five mattress planks and nail into place. Invert the entire unit

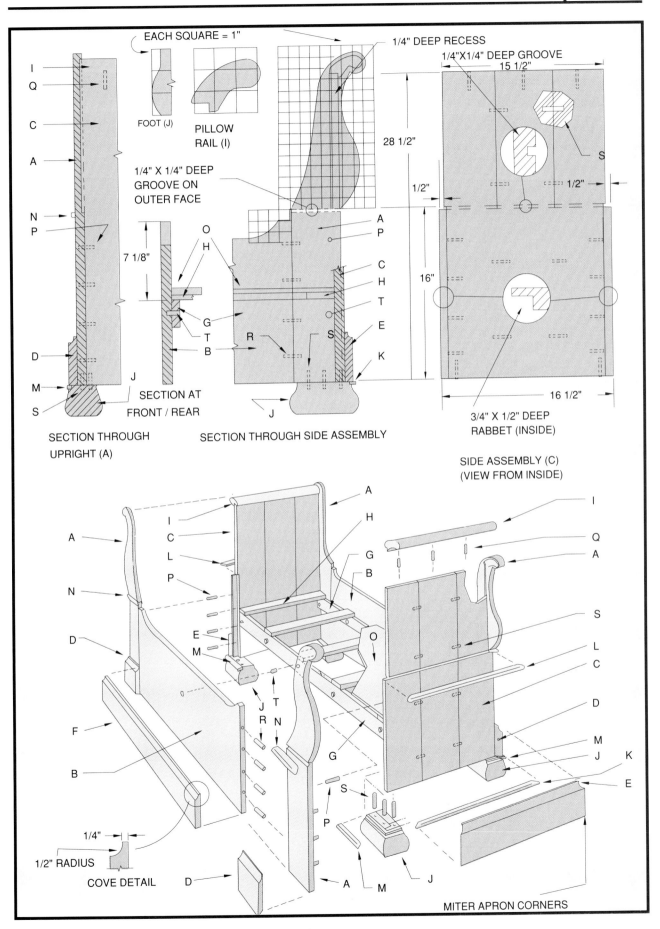

EACH SQUARE = 1"

FOOT (J)

PILLOW RAIL (I)

1/4" X 1/4" DEEP GROOVE ON OUTER FACE

1/4" DEEP RECESS

1/4"X1/4" DEEP GROOVE

15 1/2"

28 1/2"

1/2"

16"

16 1/2"

3/4" X 1/2" DEEP RABBET (INSIDE)

SIDE ASSEMBLY (C)
(VIEW FROM INSIDE)

SECTION AT
FRONT / REAR

SECTION THROUGH
UPRIGHT (A)

SECTION THROUGH SIDE ASSEMBLY

7 1/8"

1/4"

1/2" RADIUS

COVE DETAIL

MITER APRON CORNERS

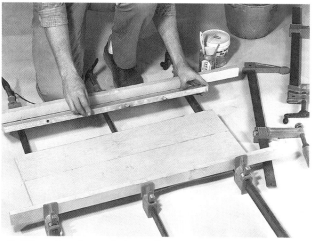

1 *Edge-glue pieces of pine to form the side panels. Secure the pieces with bar clamps and apply light pressure.*

2 *Cut out the foot design on a band saw, supporting the workpiece with the waste from the initial cut, as shown.*

3 *Stack and nail four boards together. Then lay out the corner upright design with a paper template and pencil. Gang saw all four corner pieces at one time. Use a sharp blade.*

4 *Shape the rail before cutting it to length. Jack and gutter planes were used for shaping the contours. Sand smooth.*

onto sawhorses, resting it on the planks. The sawhorses should be high enough so that the upright corners clear the floor.

To make the apron, rout the cove along the top edge, using a table saw mounted with a moulding head cutter. Sand the coved board, then cut the mitered apron pieces as shown in the drawing. Glue the apron to the cabinet.

Cut the beaded pieces and round

BILL OF MATERIALS — Victorian Fireplace Bench

Finished Dimensions in Inches

A	Corner Upright	$3/4$ x $7\frac{1}{2}$ x 31 pine	4
B	Front/Rear Panel	$3/4$ x 15 x 39 pine	2
C	Side Panel	$3/4$ x $16\frac{1}{2}$ x $28\frac{1}{2}$ pine	2
D	Corner Apron	$3/4$ x $4\frac{1}{2}$ x $5\frac{1}{2}$ pine	4
E	Side Apron	$3/4$ x $4\frac{1}{2}$ x 18 pine	2
F	Front/Rear Apron	$3/4$ x $4\frac{1}{2}$ x 39 pine	2
G	Slat Support Rail	$3/4$ x $2\frac{1}{2}$ x 47 pine	2
H	Mattress Plank	$3/4$ x $2\frac{1}{2}$ x 15 pine	5
I	Pillow Rail	$1\frac{1}{2}$ x $3\frac{1}{2}$ x $15\frac{1}{2}$ pine	2
J	Foot	3 x $3\frac{1}{2}$ x $6\frac{1}{4}$ pine	4
K	Side Bottom Bead	$1/4$ x $1/2$ x $18\frac{1}{2}$ pine	2
L	Side Upper Bead	$1/4$ x $1/2$ x 17 pine	2
M	Front Bottom Bead	$1/4$ x $1/2$ x $5\frac{3}{4}$ pine	4
N	Front Upper Bead	$1/4$ x $1/2$ x $4\frac{3}{4}$ pine	4
O	Mattress Board	$3/8$ x 15 x 47 plywood	1
P	Dowel	$1/4$ dia. x $1\frac{1}{2}$	16
Q	Dowel	$3/8$ dia. x $1\frac{1}{2}$	6
R	Dowel	$3/8$ dia. x $1\frac{3}{4}$	16
S	Dowel	$3/8$ dia. x 2	28
T	Dowel	$1/2$ dia. x 2	10

5 *With a pin router, rout grooves to fit the end panels in the corner uprights. Rout freehand by cutting next to the line, if desired.*

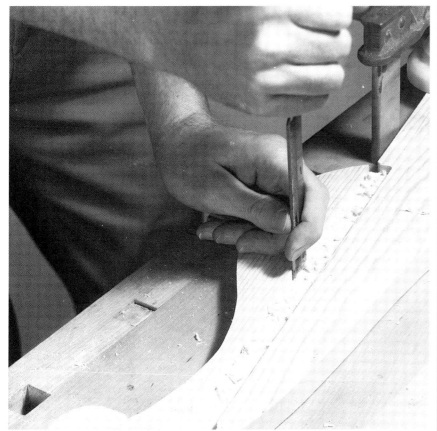

6 *Cut the groove slightly undersize and dress with a hand chisel to insure that dimensions are exact.*

over one edge with a small plane; sand the pieces smooth with fine sandpaper. Cut, miter and fit the pieces into the corner and end panel grooves.

Bore three holes for the dowels in the feet. Two dowels will enter the corner upright and one will tie to the end panel. Locate matching holes with dowel centers, bore and insert dowels. Glue and fit the feet to the unit. Cut and miter beaded pieces to fit the grooves in the feet and across the underside of the panels. Glue and nail into place.

Next, tip the unit upright, and cut and fit the plywood mattress board on top of the mattress planks. Lift the unit onto sawhorses, resting it on the planks. Sand the entire unit with 180 grit paper, and wipe away all dust with a tack cloth.

Then apply a coat of Minwax Ipswich pine, allowing it to penetrate before wiping off the excess. Let it dry overnight. Using No. 0000 steel wool, polish in two or more coats of orange shellac.

Cover the 6 in. foam cushion with a medium-weight upholstery fabric, sewing it like a fitted sheet. A color that picks up the tones of the curtains in the room was used in the sample bench. — *by William Beyer and David Warren.*

7 *Rout a dado into the side panels (shown) and the corner uprights to accommodate the decorative beading.*

Microwave Cabinet

Finding a spot for the microwave is always a problem.
This storage cart on wheels provides a very flexible solution.

More than just a microwave cart, this cabinet unit provides an area for storing hot pads, utensils, microwave dishes and other supplies. It even features a pull-out shelf for serving or preparing food.

Tips

Use your imagination when finishing the cabinet. You might consider bright-colored paint, wood stain or even plastic laminate. Select only high quality ¾ in. plywood, sanded on both sides.

Apply ¾ in. iron-on edging to the project's showing edges, particularly the pull-out shelf, drawers and doors.

Construction

Begin by cutting the top, back, side panels and bottom to their proper length and width. Make straight cuts using a circular saw guided by a straightedge. Also use a plywood cutting blade.

Apply ¾ in. wide edging to the showing edges of the side panels. Then joint the back edge until you end up with the necessary width.

Cut all of the cleats to their lengths and widths, and secure to the side panels with carpenter's glue and 2d finishing nails. Then secure the sides to the top and the bottom with glue and 6d finishing nails. Install the back panel to square the unit.

This project is courtesy of Georgia-Pacific Corporation.

149

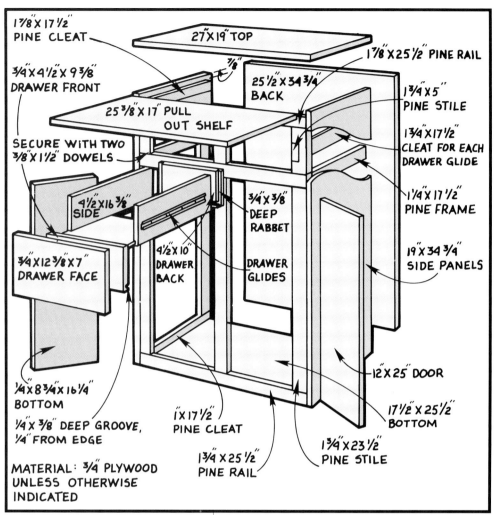

1⅞" X 17½" PINE CLEAT

¾" X 4½" X 9⅜" DRAWER FRONT

27" X 19" TOP

⅞"

25½" X 34¾" BACK

1⅞" X 25½" PINE RAIL

1¾" X 5" PINE STILE

25⅜" X 17" PULL OUT SHELF

1¾" X 17½" CLEAT FOR EACH DRAWER GLIDE

SECURE WITH TWO ⅜" X 1½" DOWELS

4½" X 16⅜" SIDE

¾" X ⅜" DEEP RABBET

1¼" X 17½" PINE FRAME

¾" X 12⅜" X 7" DRAWER FACE

4½" X 10" DRAWER BACK

DRAWER GLIDES

19" X 34¾" SIDE PANELS

¼" X 8¾" X 16¼" BOTTOM

¼" X ⅜" DEEP GROOVE, ¼" FROM EDGE

12" X 25" DOOR

17½" X 25½" BOTTOM

1" X 17½" PINE CLEAT

1¾" X 23½" PINE STILE

MATERIAL: ¾" PLYWOOD UNLESS OTHERWISE INDICATED

1¾" X 25½" PINE RAIL

plywood bottom. Now cut the plywood drawer bottom on your table saw.

Dry-assemble each drawer along with the drawer bottom to insure proper fit. Then assemble with glue and 4d finishing nails. Make sure the assembly is square before you allow the glue to cure.

Mount the drawer knob to the face. Then position the face onto the drawer front as it will appear when installed. Mark the location where the fastener hardware abuts the drawer front, and either gouge out or counterbore holes for fastener heads. Secure the drawer faces to the fronts with No. 6 by 1¼ in. flathead wood screws (countersunk) driven from inside the drawer.

Finishing

Finish sand the entire project, making sure that all irregularities are filled in with wood filler and that the edges of the project are slightly broken. Then remove dust from the project parts with a damp cloth.

Apply a sanding sealer to the project and sand lightly. Then apply two coats of your favorite finish, sanding lightly between coats.

Insert the pull-out shelf in the cabinet and add a ¾ in. by ¾ in. by 25 in. cleat to the bottom innermost edge of the pull-out shelf. This needs to be accomplished from underneath the cabinet. Secure the cleat with No. 6 by 1¼ in. flathead wood screws (countersunk). This cleat acts as a stop.

Add the knobs to the doors and hang the doors with hinges. Add two magnetic door catches for each door and install the drawer glides.

Finally, complete the unit by installing a set of four casters to the bottom of the cabinet. — *by Jonathan Wesley.*

1 *Use a doweling jig to drill the dowel holes into the cabinet stiles and rails.*

Carefully measure and cut the pine stiles and rails for the cabinet front. Dowel and glue the stiles and the rails to one another and assemble on a flat surface. After the glue has cured for 24 hours, finish sand the stiles and rails and attach to the inside of the cabinet. Assemble with glue and 6d finishing nails.

Measure the pull-out shelf cavity and construct the pull-out shelf, allowing for iron-on edging. After cutting the pull-out shelf to its proper length and width, apply the edging and finish sand.

Also cut the two plywood doors at this time and similarly apply edging to all four sides.

Remeasure the drawer cavity and cut the drawer's sides, back, front and face. Rabbet the two outside edges of the drawer back on your table saw using a dado blade. Then groove the drawer back, sides and front to accommodate the

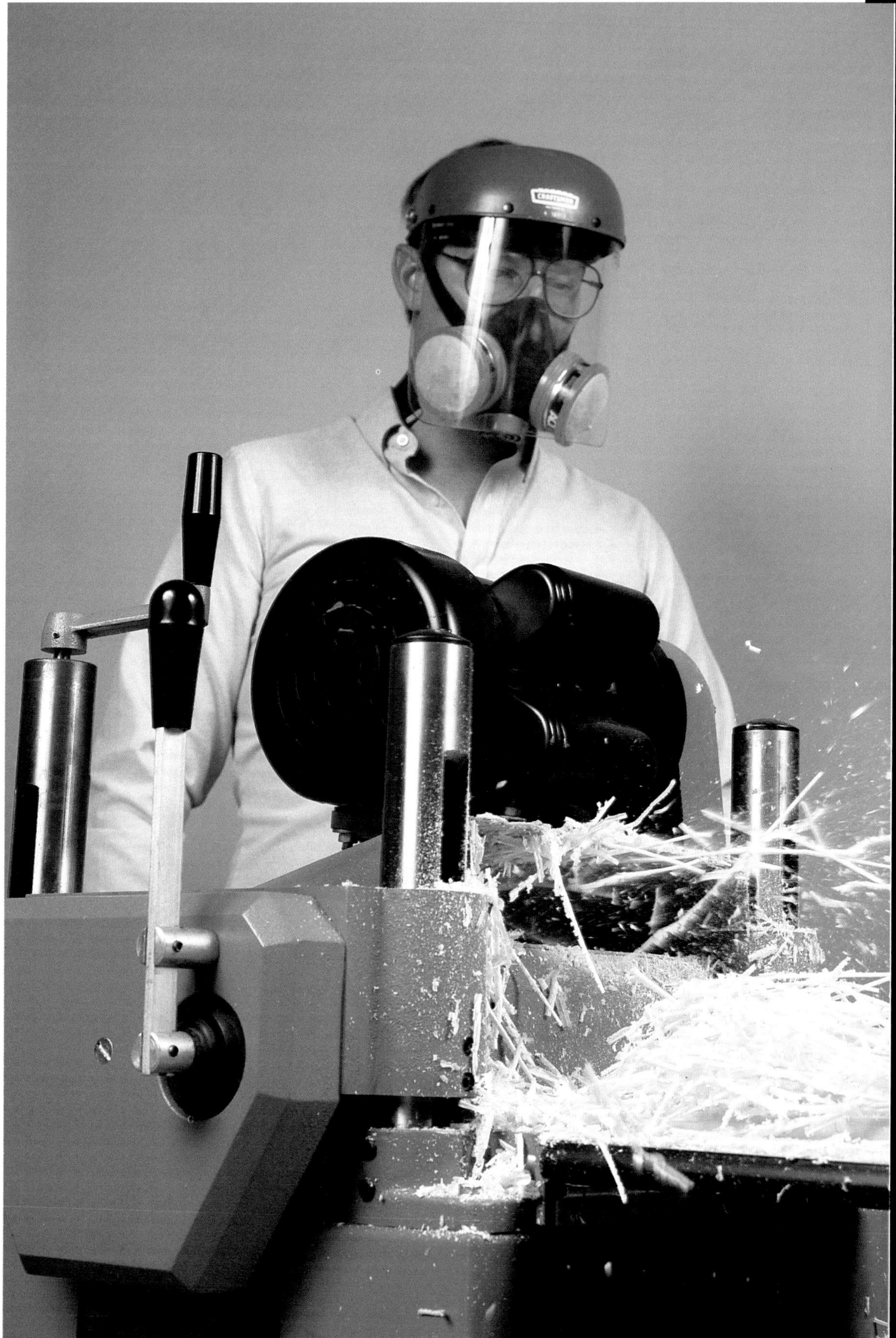

IV. Tools & Techniques

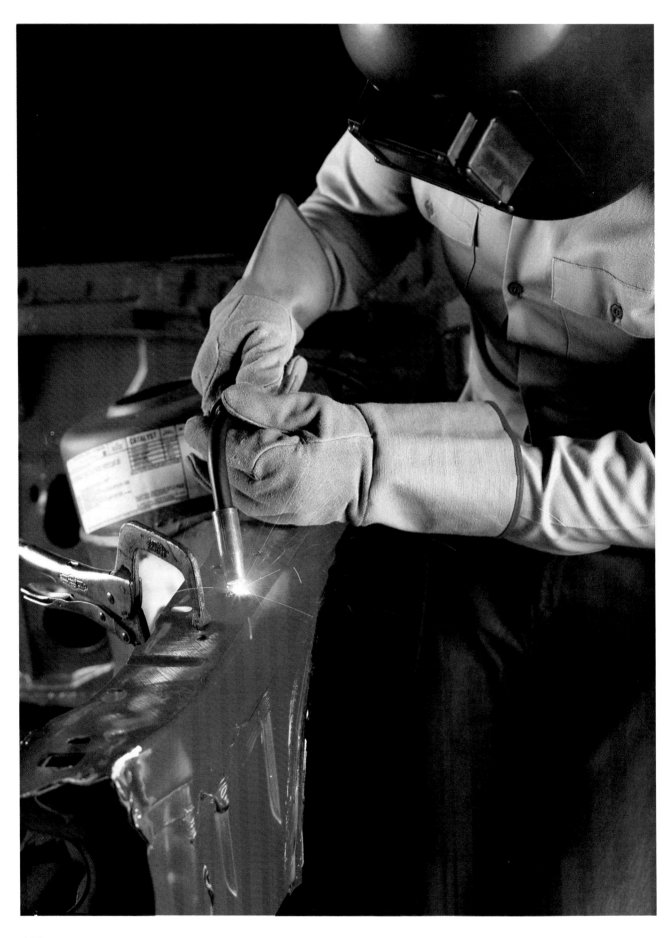

Welding For The Do-it-Yourselfer

Gasless mig welding is no longer a skill better left to the professional. Here's how you can learn to do countless metal repairs around the house.

Many homeowners think about the possibilities of welding whenever they throw away objects that could have been repaired with a good weld. Simply watching the welding process is enough to stimulate the do-it-yourselfer to try welding. However, until recently, welding has been a skill that required extensive training and expensive equipment. Nowadays, innovations in welding equipment and processes mean that homeowners can learn to weld with only a minor investment in time and equipment. Today anyone can do many of the projects that were once limited to professional welders.

The opportunity for home welding greatly improved with the manufacture of 110 and 230 volt single-phase welders (single-phase refers to the type of electrical power found in most homes).

Many types of welding processes are available, such as arc welding (gasless welding), mig welding (uses gas) and wire-fed arc welding (gasless welding).

Until recently, gasless welding was limited to certain types of welding jobs, but today, it offers many advantages over the use of mig welders. Specifically, gasless welding offers the following advantages:

- It's easy to do and requires little training.

- It can be used indoors or outdoors.

- It requires no expensive gases.

- It allows you to weld rusty metal.

Whether you choose gasless or mig welding, you'll find that home welding can be fun and money-saving. Once you get started you'll be surprised at the number of projects you can apply welding to.

Safety First

Use common sense when setting up your welding station. Remove all flammable materials from your area, make sure you have adequate ventilation and take precautions to avoid electrical shock hazards.

A welding arc can cause eye damage to even the casual observer, so do your welding in a location that is obstructed from the view of others. Welding screens are available at your local welding supply store.

For yourself, a shield with a No. 10 filter provides the minimum protection from arc flash. Most of today's welding sets for the home hobbyist come with an appropriate helmet or shield.

Wear heavy welding gloves and a close-weave cotton or leather jacket with a collar to protect your neck and arms from the burning rays that are emitted from the ultraviolet light of welding.

Needed Tools

Anytime you weld you may need to shape or hold the metal, so certain tools will be necessary. With gasless welders, there is an afterburn on your weld, called "slag," that is left over from the flux on your wire. A chipping hammer and a wire brush are effective for cleaning slag off the

1 *Today's generation of gasless wire feed welders offer the do-it-yourselfer opportunities for adding welding repair to the home shop.*

2 *A primary concern in welding is safety. Here the home welder removes common flammables such as gasoline, paint products and cleaning rags.*

This story is courtesy of Century Mfg. Co.

3 *Hanging a welding screen around your welding station will help protect others from hazards like arc flash and welding spatter.*

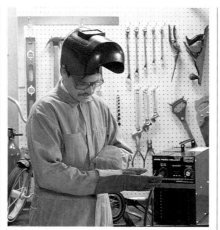

4 *The home welder's safety clothing should include items like leather welding gloves, a welding helmet with No. 10 filter, leather or tightly woven cotton coveralls and safety glasses or a face shield.*

5 *Some of the hand tools useful in welding (from left to right) are: wire snips, chipping hammer (pictured with wire brush attachment), soapstone and C-clamp.*

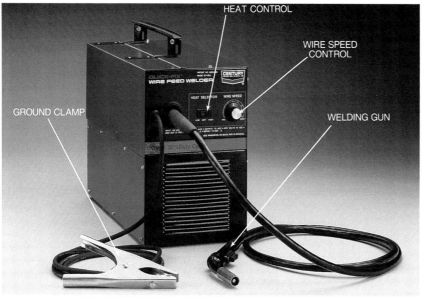

6 *Most mig welders come with controls for adjusting the amount of heat and the speed of the wire. Another control is in the gun itself, which feeds the welding wire.*

HEAT CONTROL

WIRE SPEED CONTROL

GROUND CLAMP

WELDING GUN

7 *Your welding wire spool will be either side-loaded (as shown), or top-loaded.*

welding surface. Other tools should include wire snips, pliers, Vise Grips or "locking" pliers, files and welding clamps, just to mention a few.

Preparation

First and foremost, you need to read the operator's instruction manual that comes with your welding set. This will give you all the information on the limits and capabilities of your equipment, besides telling you about how to prepare your settings, load your wire and attach your contact tips.

Once you've prepared your equipment and tested its operation according to your instruction manual, you're ready to practice gasless mig welding.

Attach your ground clamp to the welding surface or the metal bench it's resting on. Use a wire snip to trim the welding wire back until it's flush with the nozzle of the gun.

Take a couple of practice passes with your torch gun without firing it,

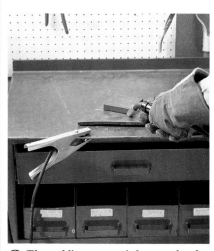

8 *The welding arc can't be completed unless the object you are welding makes electrical contact with your ground clamp, either directly or through a metal surface.*

9 *The typical parts of a mig welding gun.*

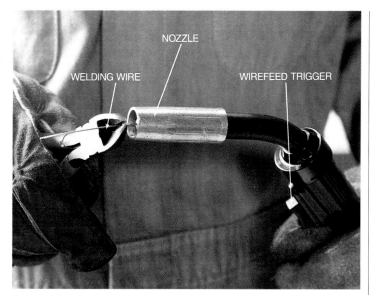

WELDING WIRE

NOZZLE

WIREFEED TRIGGER

NOZZLE ⅛″ FROM SURFACE

45°

10 *A good position for your welding gun, during the weld, is ⅛ in. from the welding surface and at a 45 degree angle to the surface.*

to get a feel for how to steady your movement. The gun needs to be held at about a 45 degree angle and about ⅛ in. off the welding surface. Here are some "don'ts" to keep in mind while practicing:

- Don't move the gun away from the welding surface.

- Don't move the gun at different speeds.

- Don't change the angle of the gun.

All of these things will affect the quality of your weld or "bead."

Start Welding

Squeeze the trigger on your gun as you allow the wire to touch the welding surface and a welding arc will start. If your machine is set correctly, you should hear a consistent buzz, similar to the sound of bacon frying. Your weld will have a low crown and little or no spatter on the surrounding surface. If you're having problems like explosions with lots of spatter, look at your bead:

- A bead with a high crown or ridge could mean that the heat was too low, the wire speed was too fast or the welding gun was held too far from the surface.

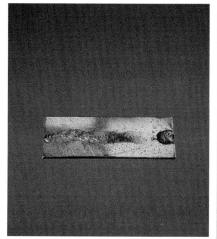

11 *The right side of the above weld shows some of the deposits left after welding with flux-coated wire. The left side shows the same weld after cleaning with a wire brush and a chipping hammer.*

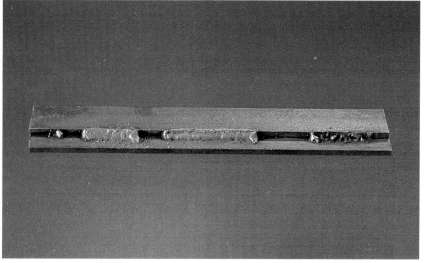

12 *Three types of welds from left to right: the first bead is incomplete and ridged, indicating low heat or fast wire speed. The second bead is correct with the metal completely bonded. The third bead appears concave and shows surrounding spatter, indicating too much heat or too slow a wire speed.*

13 *The weld pictured is a spot weld, used to hold your welding surface in place.*

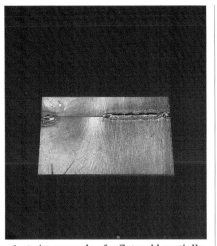

14 *An example of a flat weld partially completed.*

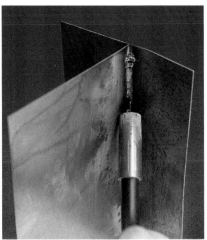

15 *An example of a vertical corner weld.*

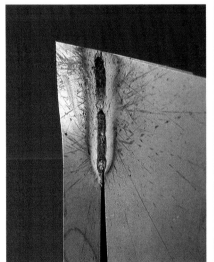

16 *Without a spot weld at each end of your welding surface, the metal may warp, making progress difficult.*

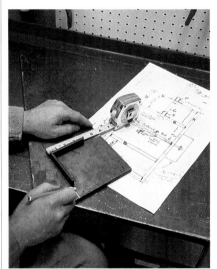

17 *The best guarantee of success in any welding project is to start with a good plan or blueprint.*

- A flat, concave bead with surrounding spatter means slow wire speed or too much heat.

Tune your controls on the equipment until you're comfortable with the bead you're laying; now you can move on to welding joints.

Although manufacturers' specifications vary, thin metals are generally regarded as those between 24 gauge and ⅛ in. thick, while thick metals are those larger than ⅛ inch. This is important information when considering how to approach your joint. Thin metal generally can be welded in a single pass, while thick metals may require several passes of the torch. Another factor that affects

your approach is whether the weld is positioned vertically versus horizontally or "flat."

Regardless of the position of the weld, the bead needs to be equally bonded to each surface or you risk a weld with little or no integrity. Start with a spot weld at each end of your joint to help keep your surfaces aligned. Metal has a tendency to warp due to the temperature change. After spot welding, begin welding where the two surfaces will join.

If you're doing a vertical weld on thin metal, keep your tip at the bottom of the weld pool and move from top to bottom. If you are doing a vertical weld on thick metal, the

process is reversed in order to make maximum use of the heat. In this case, start at the bottom and move up in a small weaving motion.

Home Project Application

There are many repairs that can be made to metal, so use your creativity and you'll discover a host of possible applications. With the right equipment, here are a few of the possibilities:

- Trailers for snowmobiles, horses, garden equipment and boats.

- Wrought-iron work for lawn furniture, fencing and stairs.

- Workbenches, engine hoists, wheel ramps and compost containers.

- Playground equipment, bicycle frames, and auto and truck frames.

- Metal art, sculptures and frames.

Like any good do-it-yourself project, a successful welding project needs a good step-by-step plan that includes the materials you need and your concept of the finished product. For additional information on wire feed welders contact Century Mfg. Co., 9231 Penn Ave., Minneapolis, MN 55431 (800) 328-2921. — *by Dan Cook. Photography by Glen Silker.*

Tools for the Home and Shop

Seven great tools that speed sawing, planing, drilling, welding...and more!

It is always exciting to search out new or novel tools. Here are some of the most exciting tools on the market for the do-it-yourselfer or woodworker. Whether you are a weekend do-it-yourselfer or an avid woodworker, we think you'll enjoy these reviews of fine tools.

The suggested retail price of each item is listed, but that price is often higher than what many retail stores charge. So shop around for the best deal.

Most are available in home centers and hardware stores, but if you cannot find the tool, write or phone the manufacturer listed. — *by Al Gutierrez.*

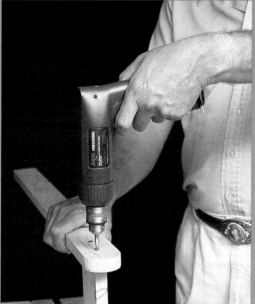

There are many affordable cordless drills on the market, but the Skil Professional (model 2503), which costs about $85, is well suited to driving screws and drilling. It features a two-speed trigger switch which moves the motor at 300 rpm for driving screws and 800 rpm for drilling. Like professional reversing drivers, it has an adjustable clutch with five settings which bury the screw to the desired depth. This tool has power and is quite suitable for home and shop use. It is made by Skil Corporation, Marketing Communications, 4300 West Peterson Ave., Chicago, IL 60646-5999, (312) 286-7330.

Delta's Sawbuck combines the versatility of the radial arm saw with the portability of the power miter box. It features an index table and track arm for cutting compound angles; a depth of cut adjustment (not calibrated) for dado cuts (accepts dado heads); and uses an 8 in. saw blade. This tool completely cuts in half a 13^1/$_2$ in. wide piece of ^{3}/$_4$ in. plywood. Best of all, the Sawbuck is portable. With the optional leg and wheel attachment, it is possible to fold the Sawbuck legs and wheel it to the work — a convenience not offered with radial arm saws. It is a high quality professional tool that will provide years of use. The Sawbuck with stand and wheels sells for $854, and is made by Delta International Machinery Corp., 246 Alpha Drive, Pittsburgh, PA 15238, (412) 963-2400.

Plane a board 6 in. wide and up to 4³/₈ in. thick in one pass with the Sears Craftsman Planer-molder (model 306.233820). For a 12 in. wide board, plane one half of the board, turn the board around and this planer-molder dresses the other half. It dresses wood at 20 ft. per minute and features a two-knife cutter head. This planer also can create moulding, with the purchase of optional cutters which fit into the cutter head. Both the planing and moulding features make the tool a bargain at about $360, stand included. It is available from the Sears Catalog. (For information on planing, read "Thickness Planer Basics" in this book.)

The Ryobi saber saw (model JSE-60) is a professional tool that rips through plywood and 2 x 4s at circular-saw speed. The saber saw's reciprocating action moves the blade's orbit into (forward) and out of (backward) the wood for a remarkably fast, rough cut. For a smoother cut, turn the control knob to a straight-line setting. Two other settings combine the speed of reciprocating action with conventional straight-line action. Adjust the blade speed to suit the cutting job, from metal to wood. An adjusting wheel varies the speed between 1,000 to 2,700 strokes per minute. After you have used this tool, you'll wonder how you got along without it. It sells for about $130, and is made by Ryobi America Corp., 1424 Pearman Dairy Rd., Anderson, SC 29625, (800) 323-4615.

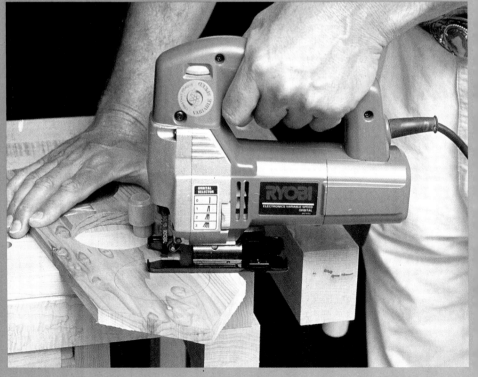

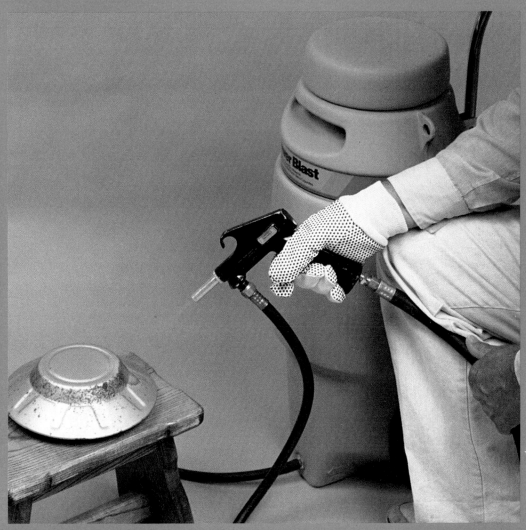

Need to sand a floor grill, bicycle or car? The Campbell Hausfeld Power Blast (model AT-1211) may be the answer. If you have an air compressor with at least 1 hp, just hook up the hose to this Power Blast to sand blast the project. It features a 15 ft. hose with gun, tungsten carbide nozzle and 9 gallon hopper. We were quite pleased with its performance. For small jobs, the 1 hp compressor is fine, but for larger jobs, like tough layers of accumulated paint, the results are more satisfying with a 2 to 3 hp compressor equipped with a 20 gallon or more tank. This unit sells for about $100, but smaller models are available. For more information, contact Campbell Hausfeld, 100 Projection Drive, Harrison, OH 45030, (513) 367-4811.

Many do-it-yourselfers are intimidated by gas and stick welders. That is why Century introduced the Quick-Fix welder. It feeds a thin wire through its nozzle as the trigger is pulled for welding. It plugs into a regular 120 volt receptacle, and features two heat settings for welding. It is quite easy to use, and very functional for a variety of home and shop applications — from bicycles to autos. The unit comes with a welding mask and spool of wire. For welding thick metals, simply make multiple passes after removing the slag. With a little practice, you will learn how to weld; conventional welding devices take much longer. Once you get the hang of it, you'll be seeking out projects to fix. The complete unit sells for $349 from Century Mfg. Co., 9231 Penn Ave. South, Minneapolis, MN 55431, (800) 328-2921. They also sell an optional kit that converts this welder to a mig welder.

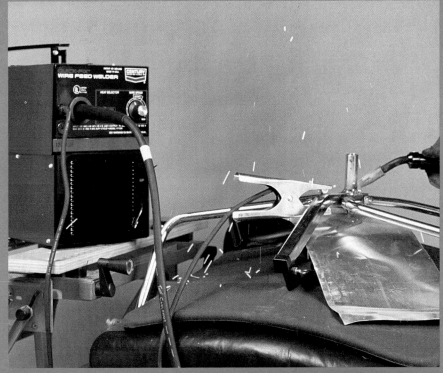

How Adhesives Solve Home Problems

Once you understand adhesives, you'll know how to successfully complete any bonding chore.

dvancements in the adhesive industry now enable homeowners to effectively complete practically all bonding chores. Regardless of an item's material make-up, its surface or color, the physical angle of the task or the natural elements to which the bond will be exposed, there is an adhesive available to solve even the most puzzling tasks.

Jim Gallagher, general manager of Devcon Consumer Division, has extensive experience in the adhesive industry and is fondly referred to as a "glue guru."

"The trick to using adhesives properly is to understand how they work," Gallagher advises. "With this knowledge, homeowners can make intelligent selections and effectively complete any bonding chore."

Almost any household chore can be completed with one of four adhesives: cements, super glues (cyanoacrylates), epoxies and acrylics (methacrylates).

Cements

Cements are popular household glues because they are fast-drying, economical and easy to use, according to Gallagher. These adhesives dry to a clear color and are effective on china, glass, wood, metal, leather and paper.

Simple mixtures of a liquid solvent and a sticky resin produce a moist, gel-like texture that makes the cements easy to apply. As the solvent evaporates, the resin clings to the surfaces, resulting in the bond.

"This type of bond is referred to as a *mechanical bond* caused by the evaporation of the solvent," Gallagher says. "Because of this, the bond will break down if it is exposed to water or extreme or continued stress."

Cements are effective on both porous and non-porous surfaces. When used on non-porous objects like glass, Gallagher advises applying a thin layer of the cement to each edge and holding or clamping the glued items until set.

For porous materials such as wood, Gallagher suggests applying the cement to each piece and allowing them to dry separately. Join the two pieces after a second coat has been applied.

Super Glues

Cyanoacrylates, better known as "super glues," are the largest selling adhesives in America.

"Cyanoacrylates are so popular because they offer a unique combination of strength, speed, ease of use and cost efficiency," Gallagher says.

"However, no matter how *super* they are, they will perform more effectively if homeowners understand how the adhesives work and how to use them properly."

Cyanoacrylates are single-component, instant adhesives that cure when exposed to the traces of moisture that are found on all surfaces. An acidic stabilizer enables cyanoacrylates to remain in a liquid state. When the adhesives are exposed to the moisture, the moisture overpowers the weak acid stabilizer and polymerization is immediately initiated on the bonding surface.

As with most adhesives, cyanoacrylates develop maximum bonding strength on surfaces that have been thoroughly cleaned and dried. It is necessary to remove all loose dirt, grease, oils and acid deposits before applying cyanoacrylates.

To test a surface for cleanliness, Gallagher suggests applying a "water break test" by running a stream of cold water over the surface.

"If the water runs off in a thin stream, it is probably clean. However, if the water beads up into drops, the surface is still contaminated and should be cleaned again with a degreasing solvent," Gallagher says.

The pH of a bonding surface also affects a cyanoacrylate bond. Acidic surfaces significantly reduce the curing speed, while basic surfaces accelerate the speed of the adhesive's chemical reaction. However, acidic surfaces can be transformed into basic surfaces. To do this, Gallagher suggests wiping the bonding area with a commercial super-glue primer or a basic substance such as ammonia.

A thin film of cyanoacrylate applied sparingly to the surface is all that is needed for a strong bond. Excessive amounts of the adhesive will slow the curing speed and result in weaker bonds. A new product on the market, the Sure Shot Super Glue Refillable Applicator, provides consumers with an easy, push-button method to accurately dispense the proper amount of the adhesive.

After the cyanoacrylate is applied, join the surfaces in the desired position and apply firm pressure

This story is courtesy of the Devcon Consumer Division.

until fixed. A good initial bond can be achieved after 10 seconds and clamps are not needed. Cyanoacrylates cure best at room temperature; however, high heat will not accelerate the process.

As little as .006 grams or milliliters per square inch is sufficient for a good bond. Because of the minute quantities needed, cyanoacrylates are extremely cost-effective in comparison to other adhesives. A 1 oz. bottle can bond 2,000 square inches of surface area.

Despite the small amounts required, cyanoacrylates provide strong bonds on a variety of surfaces. They offer tensile shear strengths of up to 3,000 psi and effectively bond metals, alloys, plastics, nylon, vinyl, urethane, acrylics, polycarbonates and rubber. However, they are not recommended for bonding glass.

The clear color of cyanoacrylates provides invisible bond lines and makes them the preferred adhesive for bonding delicate items such as figurines, jewelry, cameras and ceramics.

To remove small amounts of un-cured cyanoacrylates, Gallagher suggests wiping the area with a non-cotton cloth and then cleaning it with a solvent such as 1,1,1 trichloroethane, acetone, freon, MEK or nitromethane. *Use appropriate safety precautions and ventilate the working area when applying these solvents.* Large spills should be flushed with water and then scraped up. Cured adhesives are difficult to remove, but nitromethane is effective in thin areas.

Cyanoacrylates are activated by moisture and, because of this, Gallagher reminds homeowners that it is extremely important to tightly reseal the adhesive's container after each use. They do not need to be refrigerated, but they should be stored in cool, dry places. With proper storage, these adhesives have a shelf life of one year.

Epoxies

One of the oldest, most trusted and reliable adhesives on the market today are the epoxies, according to Gallagher. They provide tensile strengths of 2,200 psi to 2,500 psi and, because they contain no solvent that evaporates during drying, they offer gap-filling capabilities.

Epoxies are two-component adhesives that bond by chemically curing to surfaces. This reaction is produced when equal amounts of an adhesive resin and a hardening compound are properly mixed.

"The components should always be mixed with clean and, preferably, disposable tools," Gallagher suggests.

For many years, using epoxies was complicated because the components were contained in separate tubes and required careful and accurate measurement. Today, however, new chemical formulations and dual-syringe applicators have been developed by manufacturers such as Devcon to guarantee that accurate amounts of the components are dispensed.

In these new, easy-to-use epoxies, some manufacturers have added a distinctive color, such as blue, that disappears when the product is thoroughly mixed and acts as a signal that the epoxy is ready to apply.

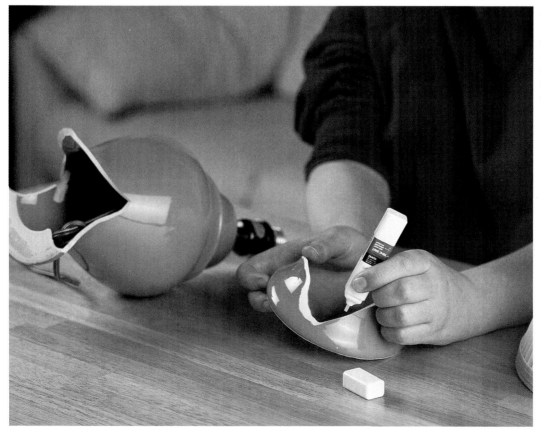

Super glue bonds instantly and is effective on surfaces such as ceramic, glass, metal and rigid plastics. A little dab goes a long way.

As with all adhesives, epoxies work best on clean surfaces. Again, Gallagher suggests running a water break test before application. He also suggests roughening the surface before application to strengthen the epoxy's bond.

"*Clean, rough* and *dry* are three key words to remember when using adhesives," Gallagher advises.

Once mixed, epoxies undergo exothermic, or heat releasing, chemical reactions. The heat resulting from these reactions acts as a catalyst that accelerates the curing process. For this reason, once the components are combined and heat starts to generate, the user has a limited amount of time to work with the adhesive.

"If the components are hot from sitting in the sun or in a warm space, the user will have less time to work with the epoxy. However, working time can be increased by storing the components in a cool place or refrigerating them prior to application," Gallagher says.

"The epoxy may be stiff to start with, but it will eventually soften and will lengthen the amount of work time."

A temperature above 50 degrees Fahrenheit is required for epoxies to cure, but if they are stored at a lower temperature they will soften and cure when the temperature rises.

Epoxies should be applied directly to one surface in an even film or bead and assembled with the mating part within the specific epoxy's working time. Firm contact between the parts for the recommended time will insure good contact and a strong bond.

"After this amount of time the bond will be at 80 percent of its strength and can be handled," Gallagher says.

Quick-setting epoxies, such as Devcon's 5 Minute Epoxy Gel, set in just three to five minutes. For this reason they are ideal for bonding items to walls, ceilings, under counters and other horizontal or vertical areas. These fast-setting epoxies are water resistant, but are not recommended for long-term immersion. They will reach full functional strength in one hour.

Epoxies with a slower setting time, such as Devcon's 2-Ton Epoxy, give the user as much as 12 minutes of working time and are recommended for bonding chores that require a precise fit. The longer setting time also results in an extremely high-strength adhesion that is more resistant to water and most chemical agents. They are suitable for bonding tasks in the bathroom and kitchen, such as mending cracked toilet bowls and tanks and replacing missing porcelain chips. Full functional strength is reached in 8 to 16 hours.

Plastic steel, another slow-setting epoxy, contains metal chips that increase the epoxy's strength and gives it an appearance that makes it appropriate for heavy-duty bonding, such as engine block repair.

"The different working times of these epoxies illustrate the importance of understanding how adhesives work before using them," Gallagher says.

Epoxies are good for bonding metals, wood, ceramics, china, glass, concrete and natural fibers. However, because an epoxy's bond is rigid, it is not recommended for most plastic items. Plastic is generally bendable and flexible and because epoxies do not give, their bonds will eventually shatter. However, they are effective for rigid plastics such as fiberglass.

Uncured, mixed epoxies can be cleaned up with 1,1,1 trichloroethane, methylene chloride or MEK. Cured adhesives can be removed with an abrasive.

Epoxies should be stored in cool, dry places. A shelf life of two years from date of purchase can be expected.

Acrylics

Acrylics, or methacrylates, are the newest, most adaptable and forgiving adhesives on the market today.

"They are a unique and significant addition to the adhesive market because they bond effectively to oily and other *as-received* surfaces," Gallagher says.

Acrylics, like epoxies, are two-component adhesives that acquire their strong and tough bond from the chemical reaction that occurs when the adhesive and activator are mixed. As with epoxies, manufacturers provide dual-syringe dispensers to facilitate the application of acrylic adhesives.

These adhesives create bonds that are flexible, waterproof and virtually temperature resistant, holding their bond in temperatures ranging from -20 to 200 degrees Fahrenheit. However, like epoxies, a temperature above 50 degrees Fahrenheit is necessary for acrylics to cure. But, in contrast to epoxies, acrylics will never cure if they are exposed to lower temperatures.

"Acrylics, being a plastic themselves, provide flexible bonds and are the most effective adhesive to use on plastic items," Gallagher advises. "In fact, one of the most popular brands of acrylics on the market today is called 'Plastic Welder.'"

Acrylics are quick-setting adhesives, allowing the user three to ten minutes of working time. For this reason, Gallagher suggests dispensing and mixing the amount that will be used within one to two minutes. The substance should be applied directly to one surface and then joined immediately with its mate. The item should not be disturbed for 15 minutes, after which time the bond will reach 80 percent strength. Full functional strength will be reached in one hour.

Acrylics can be removed and cleaned up using the same techniques and chemicals suitable for epoxies.

Cool, dry places below 70 degrees Fahrenheit are best for storing acrylics. They have a shelf life of one year from the date of purchase.

While adhesives are versatile, they can only be allowed to perform to their full ability when used properly. "When homeowners understand the dynamics of adhesives, they will know what adhesives are effective for which materials, how they should be applied and why they work. This will insure proper completion of bonding chores," Gallagher says. — *by LuAnne Sneddon and Gary Goodfriend.*

True Grits — The Facts About Sandpaper

Sandpaper is the most basic, common tool for home projects. Yet somehow it causes much confusion. Here's a review to sort out the ABCs of selecting the right sandpaper for the job.

Unless you're an expert woodworker, choosing the right sandpaper can be confusing, if not downright intimidating. Most of us head for the local home center or lumberyard and fumble through piles of sandpaper, both packaged and loose, that sport numbers like 36, 50, 150, 220, 500 and many numerals in between. Then we see "medium," "fine," and "coarse" grades, along with names such as "flint" and "aluminum oxide." Too often, the sales clerk is just as confused.

Although it would take an entire book to shed light on every aspect of wood sanding and finishing, here's a batch of basic tips from the experts at 3M Company, the most well-known manufacturer of abrasives. We asked these experts to share enough of their know-how to get you started on any wood-finishing project. As your experience grows, you will be able to fine-tune and expand into more advanced wood finishing. But for now, we'll stick to the basics.

First, not all sandpapers are alike. Sandpaper is graded according to particle size, or "grit number."

This categorizes the particles according to the number of openings per inch in a screen through which they pass. The most widely used system of grit numbers ranges from 12 (most coarse) to 600 (finest). A complete selection would include, in general, 24, 36, 40, 50, 60, 80, 100, 120, 150, 180, 220, 240, 280, 320, 400, 500 and 600. Quite an array of numerals, but only trained craftsmen use every one.

"Sandpaper," also known as "coated abrasive," is actually a misnomer. It doesn't contain sand and some do not contain paper. The various abrasives are flint, aluminum oxide and silicon carbide. Each has its own pluses and minuses. Here's a brief introduction to each type.

- **Flint:** It's the least costly, but it is slow-cutting and easily dulled. You get what you pay for, and you have little need to use flint.

- **Aluminum oxide:** The most commonly used sandpaper, it is very durable and works well when a lot of sanding is required. Also, it is usually the best bet when using coarser grades, plus it is ideal for power sanders.

- **Silicon carbide:** This is a wet/dry sandpaper that will smooth just about anything. Some woodworking experts say silicon is ideal for sanding between coats of finishes when grits of 220 and finer are needed. Waterproof silicon carbide paper is highly touted for ultra-fine sanding. Used with water, the paper flushes clean and washes away sanding waste.

Grit numbers are hashed about by professionals. Some say woodworking sanding should begin with a 100 grit sandpaper. Others advise a coarser 80. Here's some expert advice from the 3M technicians as to what the most common grades are and what they accomplish:

- **Extra Coarse (24 to 36 grit):** The best for removing thick paint or any other heavy finish. Use it for heavy rust, too. Better yet, "extra coarse" is excellent for removing deep blemishes in wood and for shaping it.

- **Coarse (40 to 60 grit):** Slightly finer, this is still a good bet for removing large amounts of wood or metal from unfinished stock. It can be a time-saver.

- **Medium (80 to 100 grit):** Here's the real workhorse for

wood finishing and perhaps the most frequently used grade. It is great for removing small amounts of wood and light rust. Some like to use it as a "starter" for sanding projects. Medium grade can be used to prepare plaster walls and ceilings for painting. But if you'll be working with drywall, try 3M Drywall Sanding Screen. Its open-mesh design resists clogging.

- **Fine (100 to 180 grit):** Here's where you begin to work your way up to the finer grits. This is for the final sanding of bare wood, plus removing slight imperfections.

- **Extra Fine (220 to 360 grit):** You'll need this for sanding between coats of finish or for producing an exceptionally smooth surface on unfinished hardwood. In fact, most furniture artisans seek such a satin-smooth finish that they go up to a 600 grit sandpaper.

Once you zero in on the grit numbers game, start thinking about your fingers. No matter what paper you go with, use a sanding block. Many pros prefer using the delicate touch of fingers when using 400 to 600 grit for ultra-smooth finishes, but most of us won't reach that pinnacle.

A sanding block adds comfort to the job, and you maintain better control as you sand. A sanding block levels the surface and helps avoid tiny humps and bumps in the sanded wood.

More tips from 3M:

- Paper is a good backing for general hand sanding, but if you use a power sander, select an abrasive with a cloth backing. It is stronger and more flexible.

- Use one of the adhesive-backed brands that adhere to the sanding block or disk.

- Sand with the grain of the wood. Circular, cross-grain

and twisting or swirling sanding strokes cause irregular cuts and scratches. These will show up in the final finish.

- Always use at least two grades of sandpaper. Start with coarse and sand the entire surface. Move up to a medium grit and sand the entire surface again. In many cases, the medium grade will be adequate for a smooth finish. For a satin-smooth finish, you may choose to follow up with a fine grade.

- Clean up between each sanding step. Use a tack cloth to remove all residue from the wood before switching to a finer grade. If you don't, loose grit from the previous grade of paper can sit under the finer paper and mar the project. Also, don't stain or finish the wood without thoroughly cleaning the sanded surface. Just as important, make sure no sanding dust is in the air when you put down your stain.

- Sand between finishing coats. This will reduce the appearance of air bubbles, brush strokes and dust.

- The easiest way to test the smoothness of the sanded area is with your forefinger. Simply move your finger lightly at right angles or diagonally across the grain (you have sanded with the grain). This way, your finger can detect small depressions or roughness. If needed, give the high spots a little extra sanding with a fine grade, but don't be too aggressive.

For additional information regarding sandpaper products contact 3M Home Products, Building 223-4S-01, 3M Center, St. Paul, MN 55144-1000. — *by Ray Lorenz. Photography by the 3M Company.*

Thickness Planer Basics

How to buy and use a thickness planer.

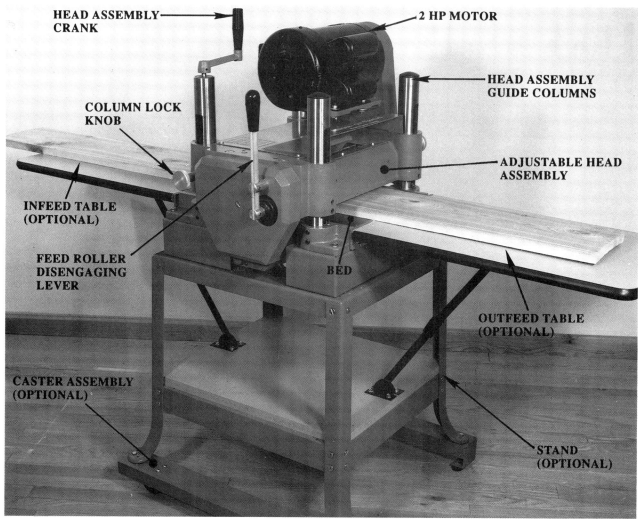

HEAD ASSEMBLY CRANK

2 HP MOTOR

HEAD ASSEMBLY GUIDE COLUMNS

COLUMN LOCK KNOB

ADJUSTABLE HEAD ASSEMBLY

INFEED TABLE (OPTIONAL)

FEED ROLLER DISENGAGING LEVER

BED

OUTFEED TABLE (OPTIONAL)

CASTER ASSEMBLY (OPTIONAL)

STAND (OPTIONAL)

Called out are the typical elements found on many thickness planers. Shown is Delta's model DC-33 13 in. thickness planer, which accepts lumber up to 6 in. thick. The tool is equipped with a 2 hp motor (220 volt, single phase), and is available with a stand with casters.

D o you know where to buy ¼ in. or ½ in. oak lumber? Locating lumber in these dimensions is difficult. That's when a thickness planer would come in handy. A thickness planer mills lumber precisely to the thickness you need. Once you feed the board into the machine, all you have to do is stand by and watch the chips fly.

Persons who do extensive woodworking or who are serious hobbyists can benefit from the operations that a thickness planer performs. Because the machine gives precisely the lumber thickness demanded, purchase of a planer by the homeowner can be a great convenience.

How It Operates

A thickness planer shaves away a board's widest surface, making it both thinner and smoother. Cutting knives, installed in the tool's head, rotate at a high speed. Rollers grab and move the lumber past the cutting knives, which literally chisel away the lumber. You control the depth of cut, and therefore the board's ultimate thickness, by turning a crank.

Technically this process is called thickness planing because the operation gives you a uniform and controlled thickness. Undoubtedly, you've heard of the jointer referred to as a jointer/planer. Like the thickness planer, the jointer/planer shaves the widest part of a board with the aid of hold-down guides. Unlike the thickness planer, the jointer/planer lacks rollers to allow controlled thicknessing.

Cutting Capacities

All thickness planers have a maximum board width and thickness they can accommodate. Thickness planers are identified by the board width they will accept. Thus, a 12 in. planer accepts a board up to 12 in. in width. However, all 12 in. thickness planers do not necessarily accept the same board thickness.

The amount of wood you can remove with the tool in one pass (depth of cut) is dependent on the

This Sears model 306.233820 planer-molder features an open side that allows you to double the 6 in. cutting width to 12 inches. If you are surfacing a board wider than 6 in., pass the lumber through the planer, turn the lumber end-for-end and surface the remaining wood. The tool also allows you to shape with special moulding cutters. The stand is included.

Ryobi's AP-10 thickness planer is a light duty machine that weighs only 57 lbs. and is designed to operate on your workbench. It comes with infeed and outfeed roller extensions.

type and width of wood you are thicknessing, the tool's motor and the design of the tool. Therefore, carefully refer to the operator's manual when cutting woods of various types and widths. The obvious rule of thumb is that as you cut narrower and softer woods (i.e. 1 x 2 pine) you approach the tool's recommended maximum depth of cut. Conversely, the harder and wider the board, the shallower the depth of cut.

All 12 in. thickness planers do not have the same recommended depth of cut capabilities. The tool is intentionally designed to meet but not exceed this depth of cut maximum. Accordingly, tools that allow a greater depth of cut have much larger motors. (A 2 hp or more motor usually requires a 220 volt hookup.)

You can always tell when you have exceeded the recommended depth of cut for the board's hardness and width because the motor dramatically slows down or completely stops. If this happens to you, immediately turn the machine off or reduce the depth of cut by turning the crank.

Why quibble about one 12 in. thickness planer that has a maximum depth of cut of 3/32 in. versus another with a 3/16 in. depth of cut? This difference actually means you will be

able to finish thicknessing in a shorter time due to removing more material on each pass.

Other capacities that you need to familiarize yourself with are the minimum length, width and thickness of material that can be safely fed through the machine. If you don't heed the manufacturer's minimum requirements, you will set up a dangerous surfacing operation. Lumber will blow apart and be blasted outward before you can blink an eye.

Quality of Cut

The quality of wood cut is determined by how many cuts are made per linear inch of material. About 90 cuts per inch give you a surface that requires light sanding, while 200 cuts per inch require little or no sanding. Cuts per inch have more value if you visually examine lumber that has been milled from the machine.

Thickness planers vary in feed rates. One that gives you a greater feed rate (the rate at which a board

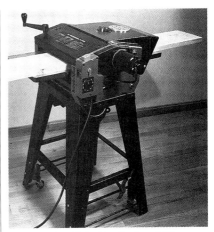

Shopsmith's 12 in. Professional Planer accepts boards up to 4 in. thick, comes with 1 3/4 hp motor (115 volt) and stand. It features variable feed rate (from 7 to 20 ft. per minute), which results in ultra smooth cuts when cut at its slowest feed rate.

passes through the machine), produces fewer cuts per inch and requires more sanding. A planer that gives you a reduced feed rate and produces more cuts per inch requires little sanding. If you do a lot of

Planer Setup

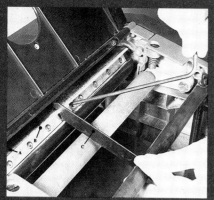

1 *After you have assembled the tool with stand, unplug the tool and check to insure that the cutting knives are set properly using the tool's supplied knife gauge (A). If a knife must be adjusted, loosen the gibs (B) and turn the knife leveling screw (C) until it meets the gauge.*

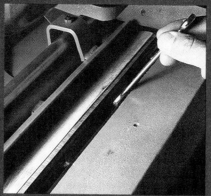

2 *Once the knife is leveled, secure the gibs by tightening the wedge locking screw or nut (shown above). Check to make sure all the gibs are firmly tightened. After every hour of use, recheck the gibs. Loose gibs create a deadly situation that can throw knives out at an explosive speed.*

3 *Lubricate all moving parts as recommended by the tool manufacturer's operator manual. Before each use, also lubricate the bed and exposed metal parts with a silicone lubricant. If you move the tool from a cold area to a warmer area, give it a heavy coat of silicone to inhibit rust.*

4 *It is likely that you will store the heavy tool in a corner of your workspace and pull it out when you need it. Thus, casters are a must. If your planer doesn't have a stand with casters, you can make your own. Construct a base consisting of two 3/4 in. plywood layers, glued together. Attach stationary casters at one end and a 2x4 leg ripped to caster height at the other.*

5 *Attach the base to the tool with L-shaped brackets. Cut two handles from 2x4s, similar to those shown above. Unplug the cord, then crank the tool to accept the handles. Insert the handles and carefully crank the head down. Move the tool by lifting up on the handles.*

cabinetry, then buying a thickness planer that gives you a reduced feed rate and more cuts per inch makes sense.

Other Benefits

Over time, a thickness planer may pay for itself because of reduced lumber costs and elimination of special setup charges at lumberyards.

If you live where lumber is in great supply, like the Midwest, large sawmills that cut and dry lumber may sell rough-cut lumber to you at substantially reduced prices. Usu-ally, these businesses require a minimum purchase of 100 or more board feet of lumber. Check with your county agent or home extension office to obtain a list of sawmill operators. Also, some larger lumberyards that do their own thickness planing will sell unsurfaced lumber

Basic Surfacing Procedure

1 *When surfacing long boards, use a roller stand on both the infeed and outfeed ends of the tool. Keep the board as horizontal as possible. Otherwise, the weight of the board will lift up the tensioned feed rollers, thereby lifting the board into the cutting head and creating a severe gouge (snipe). If the board is very heavy or long, it can generate a severe kickback or a forward thrust.*

2 *If you allow boards to fall on the outfeed end, as you grab for them your head will be positioned directly in front of the exposed cutterhead. Always avoid getting directly in front of the machine because of flying wood chips and knots. Likewise, stay away from the rear of the tool because of the likelihood of kickback. If the boards to be milled are fairly flat, then alternate surfacing, first one side and then the other, until you reach the desired thickness.*

3 *Always stand to one side of the machine when it is operating. Before the first board passes completely through the tool, butt the second board against it. This eliminates board sniping.*

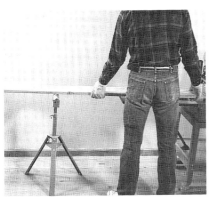

4 *Quickly grab the first board after it has passed at least 1 ft. in front of the tool. Grab the edge nearest you and push the board forward on the rollers. Pull the board toward you and then stack the lumber.*

5 *Stack all the lumber from the first pass in a neat pile (B) in front and to the right of pile A. Stack all cut surfaces in the same direction; i.e. fresh-cut surface up.*

6 *After feeding the last board, reset the depth of cut and refeed the last board, resurfacing the other side (second pass). Stack the lumber from the second pass (C) next to pile B. For remaining passes, alternate stacking between piles B and C.*

Tips For Safe Operation

1 *Feed boards that are cupped with the curved edges up, as shown. Mill this surface until it is almost (or nearly) flat, then surface the other side. Severely cupped lumber will actually give you very little remaining thickness.*

FEED DIRECTION

3 *Always mill with the grain, not against it. Shown above is the edge-grain pattern for oak. The pattern should point away from the tool. This reduces wood tear, makes the tool work more efficiently and gives you a smoother finish.*

5 *Always make shallow passes on knotty boards. Take off no more than $1/16$ in. on any one pass. Again, remove any loose knots with a hammer.*

6 *If you work in an enclosed setting, like a basement workshop, a dust collection system is a must. Not all thickness planers have dust collection chutes. Use a vacuum system that has a capacity of at least 16 gallons. Wood chips collect quickly.*

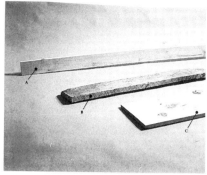

2 *Do not surface lumber that is severely bowed (A), twisted (B) or knotted (C). Bowed and twisted lumber has little woodworking value. Knotted lumber will chip blades. Always remove loose knots. They can become lethal projectiles.*

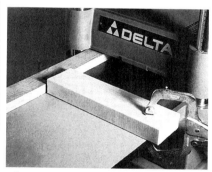

4 *Use a straightedge guide on narrow boards; otherwise they will have a tendency to feed in at an angle. With shorter boards, the workpiece can bind. If this happens to you, stop the machine and disengage the feed roller if it has a disengaging device.*

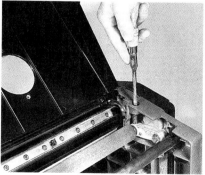

7 *If longer lumber feeds in sideways or consistently stops in the middle of surfacing, your feed rollers are probably out of adjustment. For the machine shown, the roller is lowered by turning down a machine screw at each end of the roller.*

to you at costs below surfaced lumber.

When you need lumber that is thinner but not equal to ¾ in. or 1½ in., then you probably will pay an additional planing cost or setup fee for having the lumber remilled at a lumberyard. If you buy a lot of lumber that has to be remilled and you live in an area where sawmills are not prevalent, then consider purchasing a thickness planer.

Buying Guidelines

There are about a dozen tool manufacturers that sell thickness planers. Buy a tool that matches your needs. Determine the width and thickness of board that you will process the most, as well as the volume. If you plan on milling less than a hundred board feet per year, then a lightweight tool makes sense.

Of course, if you thickness plane, you also have to joint the lumber. If you don't have a jointer, look at the cost benefits of buying a tool that comes with a jointing option.

Make sure that the tool you buy comes with a good user's manual. Some manuals are poor in explaining tool setup, safety and tool usage. Always ask to look at the manual before you buy the tool.

Here are some other factors to assess before you buy:

- Check your electrical needs. Do you have 220 volt capability?

- Determine your space limitations. You may not have room for a larger tool.

- Will dust collection be a problem? In an enclosed setting a dust collection adapter and vacuum are a must.

- Assess your future needs. Make sure that the tool you buy will meet your needs in the future.

- Determine the manufacturer's ability to provide you service, replacement parts and accessories.

- Finally, determine how much

you are willing to spend. Money is a problem for most of us, but you will be satisfied in the long run if you invest in a quality tool that meets your needs.

Added Features of Thickness Planers

The most common added feature is the jointer. Some have the jointer attached to the thickness planer so you can thickness or joint without readapting the tool. Other machines require that you reposition parts in order to joint (however, they take only a few minutes to set up).

Some tools have moulding features. They have slots for special profile cutters, or profile knives that can be purchased separately. Profile knives require removing the head's knives and installing the profile knives along with counterweights. Less available features are variable feeding and replaceable heads that allow thickness sanding.

Quality

As a rule of thumb, the heavier the tool and the greater the motor size, the sturdier and more costly the tool. Lightweight tools are designed to make lighter passes and are often built with sheet metal tables. Heavyweight tools have cast iron tables, higher quality bearings and so on. Always examine a tool in concert with its cost and construction.

Where can you find thickness planers? Many more lumberyards, hardware stores and home centers are now carrying these specialized tools. Also, look at ads in major woodworking magazines.

Lumber You Should and Should Not Mill

The handy person rarely turns down a free or inexpensive piece of lumber. If you plan on salvaging that plank by surfacing it, make sure you know what the consequences are. Used lumber may have nails, staples or other fasteners that will ruin your cutting knives, so carefully inspect all lumber.

If you buy lumber that has been painted, lacquered, etc., you may

also ruin your blades. Polyurethanes can be even harder than the wood you're surfacing, prematurely wearing your cutting knives. Some of these finishes will melt under the knives' high speed and gum the cutter head. In turn, the gumming will cause excess friction, generate heat and weaken the cutting knives.

Safety Tips

If you understand the correct feeding procedures, the type of lumber you can safely thickness and how to thickness with your machine, then you understand 80 percent of what you need to know about safety. The other 20 percent basically consists of how you dress. Wear tight-fitting clothes, good protective gloves, goggles or a face shield, ear plugs and a dust mask. You may look like something from outer space, but you'll still have your arms, hands, eyes, ears and lungs.

Thickness planers emit high decibel noise and spew out great volumes of chips and dust. The fine dust is harmful to your lungs and can irritate your eyes. Though all domestic woods are safe to surface, some imported woods can be deadly to inhale in dust form.

Use good sense. Never place your hands inside the machine unless the tool is unplugged. Keep the machine unplugged and locked away from children. The cutting knives on these machines are unforgiving. — *by Al Gutierrez and Home Mechanix magazine.*

Sources:
12 in. Professional Planer: Shopsmith, Inc., 3931 Image Dr., Dayton, OH 45414.

Delta DC-33 Planer: Delta International Machinery Corp., 246 Alpha Drive, Pittsburgh, PA 15238.

Ryobi AP-10 Planer: Ryobi America Corporation, 1424 Pearman Dairy Rd., Anderson, SC 29625, (800) 323-4615.

Sears Planer-Molder Model 306.233820: Sears Catalog.

How to Buy a Circular Saw

A guide to getting the tool that is right for you.

Cutting wood is basic to most home remodeling and workshop projects, and the circular saw is the best tool to do the job. Just walk down the power tool aisle of any hardware store or home center and you can see how popular this type of saw is. There's a tremendous variety and selection to choose from.

Obviously, the $30 bargain special is not the same as the $200 professional saw. To help you decide how much saw you really need and what kind, we've put together these circular saw basics.

Styles of Circular Saws

Circular saws come in three basic styles: top handle, drop foot and worm drive. The most popular style is the top handle, because it is available in prices ranging from bargain to expensive. The drop foot and worm drive saws are generally

How To Buy a Circular Saw

1 *The connection of the cord and the handle is a high-stress area. Choose a saw with a heavy-duty, strain-relief cord.*

2 *Check the bevel and height adjustment levers. Levers with quick-release, positive-locking cam action are preferred.*

3 *Inexpensive wing nuts used as bevel and height adjustments are difficult to lock securely and loosen easily. Stamped steel wing nuts will lose their threads under heavy use.*

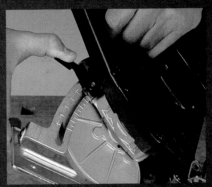

4 *Look for conveniently located, clearly marked built-in adjustment scales. This depth of cut scale is easy to read.*

5 *If you will use your saw often, make sure that the brushes can be serviced. Professional saws have external brush ports.*

6 *Homeowners will appreciate the lightweight 6¹/₂ in. saw that cuts a 2 x 4; the standard 7 ¹/₄ in. saw is most versatile.*

Saw Safety

Safe tool use is mostly common sense. First, know your tool by reading the owner's manual thoroughly.

Use only a grounded extension cord and avoid using the saw in damp, wet areas. Before you plug in any tool, check to see that it is turned off. Always remove the cord whenever you change the blade or adjust the saw. Use a sharp blade and make sure that you have installed it with the teeth facing the direction of the arrow on the blade guard.

Before using any power tools, remove jewelry and roll up shirt sleeves to prevent the tool or workpieces from getting caught on loose items. Always wear eye protection.

Do not use the saw on material it was not designed to cut, and do not push the tool beyond its capacity. Always rest the wood being cut on a sturdy surface. Support long boards to prevent the blade from binding. Never hold short pieces while cutting; clamp them securely to a larger piece. □

1 *Always use a sharp blade and install it by matching the rotation of the saw marked on the blade guard.*

2 *When changing blades, clean the area around the blade guard and check for free movement. The guard should snap closed.*

3 *For safe, efficient cutting, set the depth of your blade about 1/4 in. deeper than the stock being cut.*

4 *Always wear eye protection and support your work securely when using a circular saw. Keep the work area clean.*

5 *When ripping, use a straightedge to keep the saw from binding. Support your work and keep sawhorses and the cord from the cutting area.*

6 *Cut short pieces of stock safely by clamping them securely to a larger board. Never hold short stock in your hand while cutting.*

higher priced.

You are probably most familiar with a top handle style because all hardware and home centers carry this type. The front of the saw motor housing is secured to the forward section of its base with a hinge that allows you to adjust the depth of cut. The higher the motor housing is raised above the base plate, the shallower the cut. As the motor housing is raised, the angle between the base plate and the saw's handle increases.

The motor is mounted perpendicular to the blade for balance. This style of saw is available in a wide range of blade sizes and power ratings.

The drop foot circular saw is basically the same as the top handle, but instead of the motor housing pivoting for depth of cut adjustment, the motor housing of a drop foot moves straight up and down. This allows the saw handle to maintain a constant angle with the base plate. This feature is available only in the more expensive lines of saws.

The powerhouse of circular saws is the professional worm drive. With the motor mounted in line with the base plate, it transmits power to the blade through worm drive gearing. This gearing and the proper blade lets the saw develop the power needed to cut through metal or concrete. The worm drive has the blade on the left side so it is more visible to the right-handed user.

Matching the Saw to the Job

Whichever style of saw you choose, the most significant features to consider are blade capacity, power (amp rating), bearing construction, adjusting levers and cord quality.

Unless you plan to cut stock thicker than 1½ in. framing lumber at a 45 degree angle, a 6½ in. (blade diameter) saw will serve you well. This size is popular with carpenters because it is light and easy to handle. These qualities make it a wise choice for the weekend do-it-yourselfer.

The 7¼ in. saw is the most popular size. It cuts all framing lumber at a 45 degree angle and has the widest selection of special purpose blades. This is important if you need to cut materials other than wood.

Saws with larger blades (up to 16 in. in diameter) are available and are considered special purpose tools. Landscapers and framing carpenters use them to cut heavy timbers and railroad ties.

Compact trim saws with 5½ in. blades can cut 2 x 4 stock at 90 degrees. This type of saw can be used for general work around the house and is a good choice for anyone wanting a lightweight saw.

The most truthful indicator of a saw's power is the motor amp rating. Horsepower figures can be misleading, because either input, output, blade or developed horsepower figures can be used (and they seldom are identified).

A saw with a 10 amp or at least 2 horsepower motor is powerful enough for most cutting jobs. If you plan to use your saw for extended periods of time or to cut large quantities of heavy stock, consider a saw with more power. Higher amp rated motors run cooler, last longer under normal work loads and have the reserve power to prevent stalling and burnout when cutting heavy stock.

The bearing construction of a saw determines its duty cycle (life expectancy). Many of the popular priced saws have oil-impregnated sleeve bearings. This type of construction allows the manufacturer to produce a serviceable saw at an attractive price. If you plan to use your saw for light jobs around the house and an occasional construction project on weekends, then choose a saw with sleeve bearings. For light homeowner use, this type of saw will hold up well and provide the best value.

If you plan to use your saw regularly for a major project such as a room addition, consider a saw with needle and roller bearing construction. These bearings are designed to withstand heavy use and require little maintenance. Many manufacturers offer popular priced top handle saws with roller bearings. All professional lines of saws are equipped with roller bearings.

This story is courtesy of The Family Handyman magazine.

In order for the saw to cut accurately, the adjusting levers must lock securely. The better saws have adjustment levers for depth of cut and blade angle that lock tight but are quick and easy to release.

Many popular priced saws use wing nuts instead of levers. They are satisfactory if the wing nuts are cast and have large wings to provide the necessary leverage. However, even large nuts tend to jam and are difficult to loosen for adjustment. Some low priced saws have small stamped steel wing nuts that are difficult to tighten with your hand and to loosen once tight.

Behind the adjustment levers there should be easy-to-read scales. Every saw should have a clearly marked bevel scale. Some have additional scales to help set the depth of cut or have scales stamped along the side of the foot to help you gauge how far you have cut into a board. These scales are handy when making repetitive cuts.

The type of cord that the manufacturer supplies with the tool is a tip-off to its quality. The cord takes a beating where it enters the handle of the saw.

To prevent wear, better saws have a strain-relief sleeve that surrounds the cord and makes it take a gradual bend when pulled. The separate strain-relief is an important feature in a professional tool. If the cord becomes damaged, the pro can cut the cord down and pull the good part through the strain-relief and reconnect the cord and start working again.

Lower priced saws have strain-reliefs, but they are usually short and moulded into the cord. These cords are usually plastic, which become stiff and brittle in cold weather.

Only the top-of-the-line professional saws will have all these features. Which one does the average handyman need? In general, to find a serviceable saw, you will have to spend more than $50 (even if the saw is on sale) to purchase a 6½ in. or 7¼ in. top handle saw with at least a 10 amp or 2 horsepower motor and positive-locking levers. — *by Katie and Gene Hamilton, Steve Wolgemuth.*

Car-Care Quiz

The sources of some car problems are hard to pinpoint. Here's expert advice on how to solve three common car-repair mysteries.

Problems

1 Moans and Groans

My son's three-month-old foreign sports car had begun to draw a lot of attention, but not because of its looks. Every time the front end moved vertically it emitted a loud ghostly groan. This got my son a lot of attention, but not the kind he wanted.

The car had been back to the dealer twice. On the first visit the dealer inspected the rack-and-pinion steering system and MacPherson-strut front suspension, lubed the lower ball joints (since they are the only components with grease fittings), and aligned the wheels. The dealer said that he expected that the noise would disappear in a few days. It didn't. On the second visit the dealer replaced the front struts and sprayed the lower control arm bushing with silicone. Again, no effect. To make matters worse, the dealer couldn't schedule the car for a more thorough inspection until late the next week. He assured my son that there was no danger of the front end breaking apart, and that it was simply a matter of locating the source. By then my son's patience had run out, and he asked me for help.

As he maneuvered the sports car up the driveway's slight incline, I couldn't help but hear the annoying groan emanating from the front end. Getting down to business, I first pushed firmly down on the front bumper. Not only did I hear the groan, but I could feel rapid vibrations with the palms of my hands as the front end dove down and rose up. I opened the hood and snooped around the engine compartment while my son bounced the front end slowly. I found no obvious signs of trouble.

What about the underside of the car, I wondered. I positioned the front end on a pair of safety stands with a hydraulic jack, chocked the rear wheels and slipped beneath the car. I poked around the undercarriage, but nothing seemed to be damaged. Then I grabbed a spray can of silicone and soaked the lower control arm bushings and sway bar bushings, along with everything that bends, pivots and swivels.

Then I pumped a few shots of grease into the lower ball joints until each dust boot puffed up like a balloon. I checked for loose front-end components with some metric wrenches, but all was tight.

Now I wondered whether the noise would change after turning the tires. We lowered the car onto the ground, and I jumped into the driver's seat. After I turned the wheel hard counterclockwise, my son pushed down on the front end. The groan was now louder than ever. What about the other direction? I turned the wheel hard right. My son pressed down again, and the noise got louder once more. Now I straightened out the front tires and my son pushed down on the bumper. This time the noise sounded as it had in the beginning.

The knowledge that alternations to the steering position changed the magnitude of the noise brought up another question: What happens if one tie rod is disconnected?

I spun the steering wheel hard left to bring the right tie rod closer to the outside of the car for easier separation. I extracted the cotter pin, zipped off the 19 mm castle nut, wedged a front-end fork between the tie rod and the steering arm, and smacked the fork with a hammer until the rod separated with a loud pop. As my son bounced on the front end, the groan was now less audible. I reconnected the right tie rod and twirled the steering wheel all the way to the right. As before, I detached the left tie rod from the steering arm. Again he pushed down on the car bumper, and the noise sounded nearly the same as when I had uncoupled the right tie rod from the steering arm.

With only the right tie rod con-nected to the steering arm, I straightened the right tire with the steering wheel. My son moved the front end up and down, and the groan was even fainter. Where was that noise coming from?

2 Twinkling Blinker

Turn-signal indicators only flash when you push up or down on the turn-signal lever. Right? Well, in my 15-year-old Chevrolet the right turn-signal indicator twinkled like a star whenever the headlights or parking lamps were switched on.

The possessed turn-signal in-dicator introduced itself while I was driving late at night. At first the light glowed weakly, then quickly shut off. During that trip I scanned the instrument panel hoping to get another glimpse, but the light failed to reappear. Maybe it was a fluke, I thought.

Two weeks later the gremlin returned, only this time the light glowed intermittently, flashing at the slightest bump. I eased off the gas and decelerated the car from 60 mph to a steady 40 mph. Glancing at the problem indicator, I now ob-served that the frequency and inten-sity of the flashes had decreased. Was there a relationship between en-gine speed and the brightness of the bulb? I pulled the car onto the shoulder, placed the transmission in park and discovered that the bulb no longer flashed but stayed dimly lit while the engine idled.

Certain there was some relation-ship between engine and bulb, I watched the lamp while I revved the engine higher and higher. To my surprise, the lamp maintained its faint glow. Then, deciding to find out whether the directionals still worked, I pushed the directional level up. The flasher relay and the right directional indicator clicked and flashed in perfect synchroniza-tion. With no indication of where the problem was hiding, I drove home with a flickering green indicator light.

The following Saturday morning I prepared to attack the problem. Earlier in the week I had photo-copied my vehicle's wiring diagram in the shop manual and had high-lighted the exterior lighting circuit with a colored marker so I could dis-tinguish the lighting system's wires from the rest of the spaghetti.

As my wife worked the lights, I noticed a new twist to the mystery: The right front parking lamp was dimmer than its counterpart on the left. I wondered whether there could be a bad connection or a faulty bulb at the light assembly.

Using one hand to carefully sup-port the wobbly and weather-beaten light assembly against the bumper fitting, I unscrewed both rusted screws from the lens cover. Then I removed the lens and the old dusty bulb, and put in a fresh bulb. My wife turned on the headlights, but nothing changed. The newly in-

This story is courtesy of Popular Science magazine.

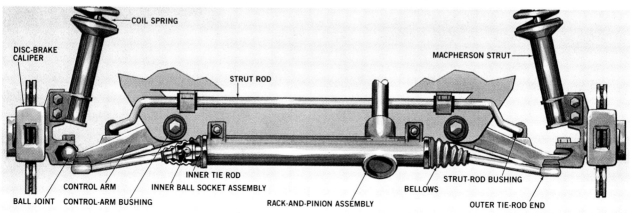

1

stalled bulb was still dimmer than its left-side companion, and the indicator bulb was glowing steadily. The problem, it seemed, was getting worse.

I took the wiring diagram under the car and tugged on the brown and dark blue wires that feed the light assembly. My wife engaged the various bulbs, but again nothing changed as I probed for frayed wires and poor connections. The wires weren't frayed and the connections were tight. Maybe the answer lies under the instrument panel, I thought, as I grabbed a flashlight, dove under the dash and repeated the same procedure on the black and dark blue wires leading to the indicator bulb. Again, no clues. What's more, there were no new developments at either light.

Now I became suspicious about the turn-signal switch itself. Maybe there is a bad connection or short circuit within the switch, I thought. I jumped into my wife's car and drove to the local tool rental store to get a special tool for removing the steering wheel.

Back home again, I pulled the stubborn steering wheel off and then carefully disconnected the suspect switch. Upon inspection, nothing seemed wrong, but because it had gone through 87,000 miles of service, I decided to retire the switch. One hour and $32 later, I'd installed a new one.

Convinced that I had solved the problem, I pushed the signal lever up, then down. Up front, my wife nodded — the directionals worked. Then I pulled out the headlight switch, anticipating another nod. But there wasn't. She shook her head in disappointment. The right front parking lamp was still dim.

I glanced over to the right directional indicator. It was still glowing defiantly. The problem remained. Where was the gremlin hiding?

3 Oil in the Air Cleaner?

"There's oil in my air cleaner!" a neighbor said to me in a disgusted tone. "It's similar to the oil-bath filter I had in my '67 VW Beetle." Before I could reply, he posed several questions rapid-fire: "How did the oil travel up to the air cleaner? Does that have anything to do with an engine idling roughly? You tuned it up three months ago. Will it need another tune-up?"

After he vented his frustrations, I managed to get him to describe what happened to his Chevrolet and its 250 cu. in. engine. "While stopped at a red light a few days ago the engine started to idle low and shake. When I got home I took off the air cleaner cover to see if the choke was open. It was open, but I found oil all over the bottom of the air cleaner."

Deciding to take a look, I chocked the front wheels, applied the parking brake, started the engine and placed the transmission in drive. The engine started to sputter and shake violently.

After turning off the ignition, I popped off the air-cleaner cover and peered inside the canister. The air filter was floating in a pool of motor oil. I poked and prodded around the engine compartment to try to find the origin of the oil. The breather element that filters air flowing into the PCV system was dripping with motor oil as I released it from its plastic cradle. This is going to be easy, I told myself. All I needed to do is to replace the breather element and the PCV valve. If the PCV valve is not open, I told my neighbor, the oil vapors and blow-by gases will accumulate under the rocker-arm cover. Partially drawn by air sucked into the engine, the oil vapors will then work their way back through the PCV system to the breather element. At the breather the heavy oil vapors accumulate within the sponge-like filter until they condense and drip into the air-filter basin.

I gave him a wink, sopped up the oil in the basin with a rag and replaced the PCV valve and the breather element. I put the throttle on fast idle, set the transmission in park and fired up the engine. As the engine warmed up, it ran smoothly and steadily. I readied a tachometer to check the idle speed.

Five minutes later, with the engine warm, a quick pull on the accelerator linkage altered the engine from a purring 1,500 rpm to a low but fairly stable 550 rpm. I placed the transmission in drive, and the smoothness disappeared. Now the engine was running so roughly that it was practically jumping off the engine mounts. And no wonder. It was idling at a mere 400 rpm — well below the 600 rpm curb-idle specification. I monitored the tachometer as I adjusted the idle speed, and finally, the bucking engine neared the 600 rpm bench mark.

With a long flat-blade screwdriver, I started to adjust the sole idle-mixture screw, which is hidden behind the vacuum lines. Keeping an eye on the tachometer and an ear to the engine, I turned the screw slowly inward. Two and a half turns later, the tach reading settled between 600 and 610 rpm. Thinking back to when I tuned up the engine three months before, I recalled that a single twist of the screw in or out produced a change of 20 to 30 rpm. Why wouldn't it change this time? I

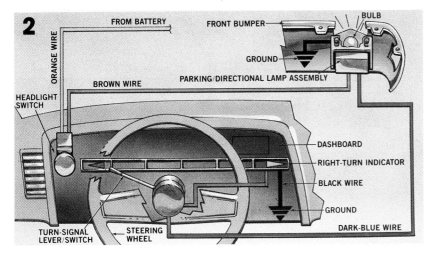

wondered. I reset the screw to various settings, but the tachometer indicator kept oscillating around 605 rpm.

Now the question was: Are dwell and ignition timing okay? Remembering that dwell changes timing but timing does not change dwell, I checked the dwell and then the ignition timing. Both parameters fell within specifications.

After I raised the idle speed and replaced the PCV valve and breather, the engine ran smoother. The problem appeared to be solved, so I put in a fresh air filter and sent my neighbor off.

Two weeks later I began to worry that I hadn't gotten to the root of my neighbor's engine problem. I walked over to his house for another look. He told me that the Chevy engine seemed to be running fine, but I convinced him that it was worth checking out one more time.

I spun the wing nut off the air cleaner and removed the cover. Once more, the air filter sat in a puddle of motor oil. What was wrong?

Solutions

1 The groan diminished each time I disconnected one of the tie rods, so I began to suspect that its source was in both inner tie-rod joints of the rack-and-pinion assembly. A close examination proved that this was indeed the case.

The noise grew when the wheels were pinned furthermost right or left because of the added pressure on the dry joints. With both tie rods disconnected and the dust boots peeled back, I could feel the dryness of the joints when each rod moved. The factory lubrication didn't last long.

Packing the dry joints with fresh EP lube and reattaching both tie rods completely eliminated my son's problem.

2 From the turn-signal indicator's behavior, I started to wonder whether something was amiss with the ground circuit of my Chevy's lighting system. Then I recalled the loose attachment of the right front parking/directional light assembly to the bumper. When I removed the worn assembly from the bumper and inspected it thoroughly, I found a crack on its mounting surface where it attached to the bumper. This made for a poor ground, which turned out to be the source of my trouble. (Vibration at higher speeds exacerbated the problem.)

The car's lighting system does not incorporate a separate ground wire. Instead, both parking and directional lights use the metal assembly that's bolted to the front bumper for its ground. When the parking lights or headlights are switched on, current flows through the brown wire into the parking-lamp filament of the dual-filament bulb. From there it exits the filament and flows through the brass base of the bulb into the metal assembly bolted on the bumper to common ground.

When the light was flickering the current sought an easier path to the ground. This easier pathway was out through the dark blue wire, all the way back to the right directional indicator bulb. The current was strong enough to make the bulb glow as it flowed out the bulb wire to ground. The wiring diagram showed that the dark blue wire was actually acting as the power feed for the right directional indicator.

What happened to my Chevy is commonly known as "feedback." Once the assembly was replaced, the right front parking lamp glowed as brightly as the left, and the right directional indicator never flickered unless I wanted it to.

3 When I saw the air filter again dripping with oil, in my mind I started working backward through the system. The oil was getting from the rocker-arm assembly to the breather. Why was this happening? The PCV system must be clogged.

If the PCV system were operating correctly, it would be sucking oil right down into the intake manifold, and no oil would be reaching the filter. I started to suspect the PCV system. I fired up the engine, pulled the PCV valve out of the rocker-arm cover and placed my thumb over the hole. Where there should have been a vacuum I felt nothing. Now I knew that there was a clog in the PCV system. Further investigation revealed there was a blockage in the vacuum/distribution tube in the intake manifold (see drawing).

Once the blockage was carefully removed, the oil problem instantly vanished and the idle speed increased 200 rpm as well. A quick readjustment of the idle speed and mixture then brought the engine completely back to normal.

One noteworthy point: After the idle speed was reduced this time,

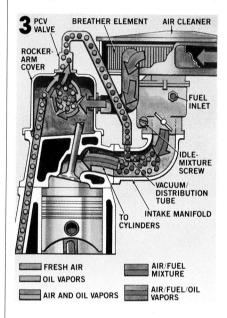

3 PCV VALVE • BREATHER ELEMENT • AIR CLEANER
ROCKER-ARM COVER
FUEL INLET
IDLE-MIXTURE SCREW
VACUUM/DISTRIBUTION TUBE
TO CYLINDERS
INTAKE MANIFOLD

FRESH AIR
OIL VAPORS
AIR AND OIL VAPORS
AIR/FUEL MIXTURE
AIR/FUEL/OIL VAPORS

turning the mixture screw had a pronounced effect on curb idle. Why? When the idle speed was adjusted prior to cleaning the PCV system, the throttle plate was opened past its normal idle position, which exposed vacuum to the fuel-transfer slot located just above the idle mixture-delivery port. The transfer slot is used to provide extra fuel for smooth engine operation while the engine is "transferring" from idle to cruising speed. Thus, turning the mixture screw while the transfer slot delivers fuel has a smaller effect on the total amount of fuel delivered to the engine at idle. It's analogous to adjusting the screw while driving nearly 20 mph. — *by Peter A. Delvecchio. Drawings by Russell von Sauers.*

Replace a Thermostat

You can save money by replacing that old thermostat.

Today, one of the easiest and most convenient ways to save on home energy bills is to use a programmable thermostat. These handy controls can be programmed to change the temperature to an energy-saving level when you're sleeping or out of the house. The technique, called offset, has been documented to save up to 35 percent on energy costs.

That's good news for people trying to keep home operating costs under control. Even better news is that many energy-saving thermostats are not only affordable, they also can be installed in less than a half hour, using only a screwdriver

and tape to code wires and terminal designations.

Programmable thermostats come in a variety of brands and styles and range in price from $50 to $120.

But they are not all alike in performance. Some have wide temperature swings which can affect comfort. Some are easier to program than others; almost all offer different programming options. Most come pre-programmed from the factory, and maintain a constant temperature until programming a specific schedule is desired. Some models can be removed from the wall after installation for easy programming.

Most programmable thermostats will pay for themselves in energy-savings within three to six months. So, invest in the best you can afford, and be sure to select a model with features that fit your needs.

For this story, Honeywell's CT3400 thermostat was selected to illustrate the installation process. This model will work on all gas or oil furnaces, with or without air conditioners.

No special wiring or relays are required because the thermostat operates independently of furnace power. The batteries will last more than a year-and-a-half; and a symbol on the LCD panel gives two or more

Remove the cover from the old thermostat. Then remove it from the base by unscrewing the three mounting screws.

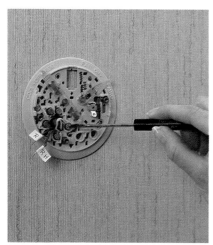

Unscrew the wires from the thermostat subbase. Be sure to label wires to correspond with the terminal from which they were removed.

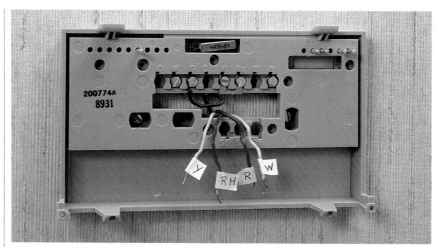

Remove the old subbase from the wall. Install the new subbase and reconnect the wires to the proper terminals.

months' warning before they need to be replaced.

The steps for installing this thermostat are typical of the steps for installing most of the do-it-yourself programmable thermostats on the market today.

Checklist: What You Should Know Before You Buy

1 Select a programmable thermostat that fits your lifestyle. Is your schedule the same every day? Then a clock-type thermostat should meet your needs. Or, if your schedule varies, then 5/2 or 5/1/1 programming is ideal.

The 5/2 designation means the thermostat can be programmed for two different energy-saving schedules: a five-day program that is the same for all five days, followed by a two-day program that is the same on those two days. Then, the 5/2 day programming cycle repeats itself.

The designation 5/1/1 means a five-day program is followed by two different one-day programs before the programming cycle repeats.

If you prefer a model with the flexibility of a different schedule for every day of the week, buy a thermostat with 7-day programming capability, such as Honeywell's Magic Stat MS3000 and CT3100.

2 Check whether the existing thermostat is a line-voltage thermostat. A good rule-of-thumb is

that if the cover can't be removed, the thermostat is most likely a line-voltage model. If you're not familiar with line-voltage thermostats, it is safest to get an electrician to install the new one.

If you are not sure, call the manufacturer of the old thermostat, or call Honeywell at 1-800-468-1502. Experienced do-it-yourselfers can purchase line-voltage thermostats, such as Honeywell's CT1600, from a local heating, ventilation and air conditioning dealer.

3 Study the label on your furnace before you buy. Note the furnace make and model. Compare this data with information on the new thermostat packaging to see if the thermostat is compatible with the old equipment. Call the thermostat manufacturer if you are not sure.

4 Determine whether the air conditioning system is a four- or five-wire system.

If you can't find the owner's manual, take the cover off the thermostat so the subbase is visible. Then, count the wires going to the thermostat.

5 Note whether the thermostat requires an "open" circuit to work properly.

Some manufacturers use electronic switches that require an open circuit. These require installing a separate isolation relay. The CT3400 and CT1500/CT1800 do not usually require an isolation relay. If

Install batteries in the new thermostat and attach the thermostat to the wall. (Be sure to use the batteries recommended by the manufacturer. The CT3400 and CT1500/CT1800 come equipped with batteries.) Then program the thermostat to fit your schedule.

the thermostat you buy requires an open circuit, contact the manufacturer for more information.

Some brands, including Honeywell and Robertshaw models, can be used on either four- or five-wire systems because they have the necessary circuits to cover most wiring contingencies. It is always smart to purchase a programmable thermostat that works with both four- or five-wire heat and cool systems.

Shop around for a thermostat. Make sure you read the thermostat's installation instructions before tackling the installation job yourself. — *by Anne Drake.*

Index